RANS[illegible]

Stories from a nature reserve and what they say about plants, conservation and us

Richard Moyse

Published by the Kent Field Club

Published by Kent Field Club, the Natural History Society of Kent.

The aims of the Kent Field Club are to promote an increased interest in natural history, to study and record the distribution and ecology of fauna and flora, and to promote nature conservation in association with the relevant organisations within the county of Kent.

First published 2023.

ISBN 978-0-956 1926-9-1

For Kathy

Contents

Sketch-map of Ranscombe Farm Reserve ii
A note on the use of names ii
Preface iii
Chapter 1 The making of the place 1
Chapter 2 The making of a nature reserve 21
Chapter 3 In the cornfields 36
Chapter 4 Arrival of the weeds 60
Chapter 5 Into the trees 79
Chapter 6 Tramps and thieves 102
Chapter 7 Building grasslands 122
Chapter 8 The gene genies... 144
Chapter 9 Spineless characters 163
Chapter 10 Putting some backbone into it 183
Chapter 11 Managing to be wild 198
Chapter 12 A place for people 219
Acknowledgements 236
References 237
Index of species mentioned in the text 247
General index 261

Sketch-map of Ranscombe Farm Reserve

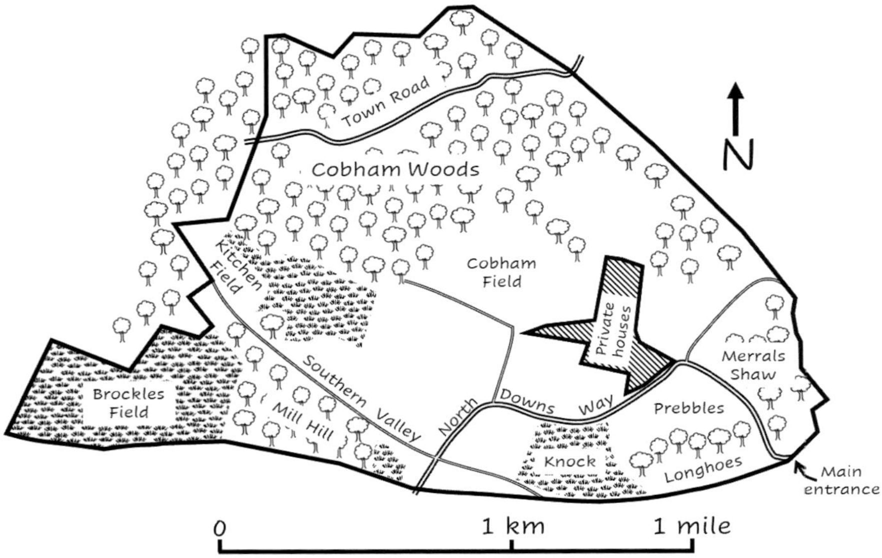

The boundary of the reserve is shown as a bold line. The map shows some of the locations mentioned by name in the text. Many smaller details are omitted, and only a small selection of the many footpaths is shown.

A note on the use of names

I have followed the convention of presenting the English names of species in lower case (except where the name of a person or place forms part of the name of a species). Scientific names are given, in most cases, on the first mention of a species. I have tried to use the most up-to-date accepted names, both English and scientific, and for vascular plants I have followed the fourth edition of Stace's New Flora of the British Isles. For bees, I have followed Field Guide to the Bees of Great Britain and Ireland by Falk & Lewington, 2015.

Preface

"On the 15th May we went to Ranscombe Farm. When we got there Mr Palmer was spraying a barley field with weed killer. They also grow potatoes, spring cabbage, cherrys [sic] and apples. There has been a farm on the site for thousands of years. The farm area is 200 acres. The first thing we looked at was the oast house. It isn't the original one because that was destroyed by a doodle bug in W.W.2. The oast house is now used for storing grain. There is a special machine there for cleaning the grain. There are 8 buildings on the farm. They changed from animals to crops in 1950. 1950 was also the best year for hops. Hops were last grown in 1959. There [is] a trapdoor in a floor of the oast house. It used to be used for pressing hops. There are marks in the oast house where fire places were. A sack of hops is called a pocket. Coal and charcoal was used to dry hops. The farm got water from a 90ft well and the water was pulled up by horse.

"A lot of modern machinery is used nowadays like the tractor the corn drier and the spray. Fertilizer is added to the soil just after planting the seed. Unlike other vegetables potatoes are planted in risen rows."

Report on a school trip made in 1974, when the author was 10 years old. Found amongst the things my parents had stored away in the loft.

In 2011, I left my previous employment and started a new job. For nine years, I had been at Kent Wildlife Trust, where I was the head of their conservation and policy team, dealing with things like planning and development, marine conservation, and work with farmers and other land-owners. The staff and volunteer team at Kent Wildlife Trust were a wonderful group of people, and I had been lucky enough to work with them at a time of real expansion in the quality and quantity of work carried out, so leaving was a wrench. But two things drove me. I had got into nature conservation because of my interest in wildlife: I liked to look at plants and animals, identify them, understand them, find things I'd never seen before. As the team I led grew and expanded during my time at the Trust, I found that my time and energies were increasingly directed to managing people, and I began to realise that getting better at my job would mean thinking less about the details of natural history and more about getting the best out of people. I take my hat off to good managers – it's a real skill and needs thought and training and care – but it was not what I got into conservation for. The other big factor was that my father was in his late seventies, living alone, and growing increasingly ill with congestive heart failure, which meant that I needed to be more available for him. My new job would be almost literally on my father's doorstep, so that I would be able to give him more of my time and be in a much better position to pop in when needed.

My new post was with another conservation charity, Plantlife, which works nationally – and to some extent internationally – to conserve wild plants and fungi. For nearly a decade, I would be Project Manager at Plantlife's flagship nature reserve, Ranscombe Farm Reserve, where I would work as part of a small, local team. Ranscombe Farm's 'flagship' status is down in part to its size: at 260 hectares (640 acres) in total area, it is the second largest of the twenty-three nature reserves owned or managed by Plantlife. Ranscombe Farm is also easy to get to, being slap-bang against the Medway Towns in North Kent, 30 miles east of London, and close to major transport links, so that it is fairly easy to reach from some of the country's larger population centres. Most importantly, it is one of the great places in Britain for wild plants, being easily the best place in the country to see cornfield weeds – the rare and vanishing flora associated with arable farmland – as well as being supplied with an interesting range of other rare plants of woodland and grassland. And, for me, coming to work at Ranscombe was coming home, because it was one of the places where I first cut my teeth as a teenage botanist and wildlife-watcher.

Immediately south of Ranscombe Farm Reserve is the village of Cuxton. It was here, in 1965, that my parents bought a house in the newly built estate which had been constructed on the south-facing slope above the old village in the valley bottom. The new estate's growth came to a halt where it met the main railway line running from London Victoria to Chatham and the Kent coast; on the other side of the railway, and just a few hundred yards' walk away from our house, were the woods and fields that would one day become Ranscombe Farm Reserve. I was two years old when we moved to Cuxton, and the house was my home till I left for university; it remained my parents' home for the rest of their lives. So the fields and paths and woods I was returning to as Plantlife's Ranscombe Farm Project Manager were the same that I used to explore as a boy with my copy of Fitter, Fitter and Blamey (many will remember fondly that wonderful Collins field-guide to wild flowers), sometimes perhaps also with binoculars around my neck, as I tried to identify plants, looked out for birds, or hoped to see a deer.

But this isn't my story, though I'm afraid you'll have to hear a bit about me along the way. What I really want to do here is to try and paint a picture of Ranscombe Farm Reserve, or the land that became Ranscombe Farm

Reserve, and to describe some of the plants and animals that make this nature reserve so very extraordinary. And it certainly is extraordinary: its one square-mile area must be one of the richest square-miles in Kent, and exceptional in the range of rare and interesting plants and animals it supports. But perhaps more than anything, the time I spent with Ranscombe Farm Reserve over the last five decades, and particularly the nine-and-a-half years I worked there, made me increasingly aware that every part of the place had some sort of story attached to it: the plants, the animals, the layout of the fields and woods, the shape of the whole landscape. It may be that, if you look closely enough for long enough at any particular patch of ground, you can find that there are stories to tell, but Ranscombe Farm provides an anthology of tales about people, places and plants that is richer, deeper and more varied than anywhere else I know.

Some of these stories tell us something about what is happening to wildlife in Britain, and how it is responding to the increasing pressures we are placing upon it. There are some stories which tell of the interconnectedness of nature. Other stories tell us the history of some of our most valued wild plants, and the history of ourselves and our relationship with nature. In places, it's even possible to draw some inferences about the future management of the wider countryside and the conservation of wildlife. I'm hoping to use this book to tell some of these stories as they appear to me – this is a personal account of a place as much as it is anything else – and to try, if I can, to keep it interesting on the way. I hope, too, that I can get across, at least a little bit, why I think the place is so special – unique, even. If nothing else, I hope I can convince you of the astonishing richness of Ranscombe Farm Reserve, why it is so important for wildlife, and why it is so worthy of appreciation and protection.

Common poppies in the Southern Valley. Photograph: R I Moyse/Plantlife

Chapter 1 - The making of the place

To start these stories, we need some background, and we need to dig down into the geological past. Ranscombe Farm Reserve is perched on the great chalk ridge of the North Downs, and lies on the western edge of the Medway Gap – the great slot through the Downs carved by the River Medway. To the north and east of Ranscombe Farm is Kent's largest conurbation, the Medway Towns, which clutches the river as it winds and broadens on its way to join the Thames Estuary. Kent itself is an interesting county. Sometimes Kent is seen as a series of dormitory towns and villages set out along the transport corridors to the continent, and this would not be an entirely inaccurate view. After all, the county is positioned neatly between London and the narrowest parts of the English Channel and it is from London that the major roads and railways radiate: a significant rail route and the M2 motorway both lead through Kent to Dover for the ferry port; almost parallel and to the south of these, the M20 and a high-speed railway run to the channel tunnel at Folkestone. Yet the county still retains a very rural and intricate landscape: compared to many parts of Britain, Kent is well wooded, and both villages and fields often remain relatively small.

That the county remains so intricate and varied is all due to the influence of geology. Kent sits on top of a huge structure that geologists call the Weald-Artois anticline. The rocks that make up this structure were laid down between around 145 million years ago and 65 million years ago, during the late Jurassic and Cretaceous Periods, when dinosaurs were everywhere. Kent's oldest rocks were laid down in a place dominated by flat coastal swamps and estuaries, and at a time when species like *Stegosaurus* were walking around. As millions of years went by, the land was slowly submerged to form a shallow sea where silts and sands were washed down from rivers and formed layers across the sea-bed. Eventually, this input of sediment stopped, and for an unimaginably long period the only material falling onto the floor of the sea was a constant rain of the tiny shells of microscopically-small, single-celled marine plankton. These minute creatures lived in the sunny upper layers of the ocean, and when they died, their shells drifted down to the sea floor where the accumulated material was eventually compacted to form the chalk we see today. The White Cliffs of Dover stand well over 100m high, a testament to the power of the accumulation of billions of tiny actions.

The last layers of chalk were laid down towards 65 million years ago – the time of 'classic' dinosaurs such as *Triceratops* and *Tyrannosaurus*. This was at the end of the Cretaceous period (*creta* is the Latin word for chalk),

View of the Southern Valley at Ranscombe Farm Reserve, looking roughly eastward towards the ridge of the North Downs beyond the River Medway.

which means, of course, that Steven Spielberg's dinosaur film should perhaps have been called Cretaceous Park. Feel free to annoy your friends with that bit of pedantry. A couple of big things happened to the planet at the end of the Cretaceous period: earth was struck by an asteroid that may have been as much as 9 miles across, resulting in catastrophic, global changes that wiped out the dinosaurs (all except the branch that evolved into modern-day birds); and there were massive movements in the great tectonic plates that form the Earth's crust. In our part of the world, Africa, India and Eurasia came together in a collision that would buckle the planet's surface and throw up a series of mountain ranges that includes the Himalayas, the Alps and the Pyrenees. As the effects of this continental crash spread out across the surface of the planet, like ripples made by a stone dropped into a pond, they became smaller and smaller in height. Over at one edge, this resulted in a series of Jurassic and Cretaceous rocks being pushed up to form a huge, rounded dome: this was the Weald-Artois anticline, a single, massive, mighty lump that stretched from modern-day Surrey and Sussex to Artois in Northern France.

The anticline was initially one enormous hill, until the processes of erosion got to work. The rain that fell on the dome started to form streams and rivers, and the rivers began to cut down into the soft surface chalk. When the rivers got down to the underlying clays, erosion could go even faster, only being resisted by the harder sandstones. This effectively chopped the top off the dome, like taking the top off a boiled egg, exposing the Jurassic and early Cretaceous clays and sandstones beneath the chalk. As the running water ate down into the softer clays, it left a ridge of chalk in a ring around the margins – the shell on the egg, if you like This ridge was steeply sloping on the inward face where it had been cut by water (this is called the scarp slope), but more shallowly sloping on its outer surface where it still roughly followed the outline of the original dome (the dip slope). Finally, the whole structure was split into two pieces by the formation of the English Channel. In what is now Kent and Surrey, the northern ridge of chalk is the North Downs, meeting the sea at the White Cliffs of Dover. The southern rim of the bowl is the South Downs, meeting the sea at Beachy Head and Seven Sisters. The two ridges are roughly mirrors of each other: in the North Downs, the steep scarp slopes face roughly south, with the shallow dip slope facing roughly northwards; in the South Downs, the opposite is true. Between the North and South Downs lies the Weald: the flatter Low Weald where H. E. Bates set his stories of the Larkin family in The Darling Buds of May; and the sandstone-capped hills of the High Weald, home of that mythical writer of letters to the press, Disgusted of Tunbridge Wells.

The North Downs are made of chalk, but that doesn't mean it forms the surface geology everywhere. Chalk is always most obvious on the steep scarp slopes where the surface is most readily worn away and the soil is thin, as well as on the sides of the many shallow valleys that ripple the dip slope of the Downs. In places, the chalk is overlain by patches of sands or gravels: these were the last deposits in the shallow, prehistoric, coastal waters before the chalk was pushed up to form the Weald-Artois anticline. On many of the flatter ridge-tops between valleys is evidence of a much later process of weathering of the surface chalk: where the chalk wasn't cut and washed away by rivers and streams, it could still be broken down by rain, snow, and frost. Over millions of years, and several Ice Ages, this slow weathering dissolved and broke down the surface chalk, and what it left behind was a thick blanket of clay, blended with a very generous helping of flints. Flint is a hard, black form of quartz which was already present as blocks and lumps and layers within the chalk, and it could resist the effects of weathering. This mixed material – clay-with-flints – dominates much of the high, level ground on the North Downs, forming soils which continue to resist plough, harrow and spade.

This, then, is the geological background to Ranscombe Farm Reserve. Sitting where it does on the edge of the gap through the Downs cut by the River Medway, the reserve overall has a roughly southerly-facing aspect, and has steep chalky slopes both above the river and on the sides of the shallow valleys that once fed water into it. More level parts of the reserve are thickly coated with the wet clay-with-flints, while the highest part of the site is on dry gravel and sand. Below Ranscombe Farm, the River Medway runs through its broad-bottomed cut, fringed by marshes; to each side, the Downs roll onwards to west and east. To the north of the reserve is the broad coastal plain and the estuaries of the Thames and Medway with their saltmarshes and mudflats. This geology, and this setting, have together come to shape the wild plants and animals of the reserve.

They've also shaped the Medway Towns themselves. It was the establishment of Rochester that marked the Towns' starting point, and this happened because the invading Romans, in 43AD, had to find a suitably narrow point for a permanent crossing of the river. The River Medway was a significant barrier to their progress, and it had taken the Romans two days of fighting against the resisting Britons before they managed to get across the Medway in the first place. The spot the Romans chose for their crossing is still the location of the modern-day road-bridge which links Rochester to Strood. Rochester was therefore initially established as a well-defended point on the road which led up to and beyond London from the Romans' bridgehead near Sandwich on the East Kent coast. This road is now the A2, but probably better known as Watling Street, the major Roman highway which extends beyond the City of London through St Albans and onwards to Wroxeter in Shropshire. Rochester High Street still follows the line of the Roman Road, and sections of the road itself have turned up beneath the cellar floors of High Street shops.

Rochester eventually became a substantial, walled city, with a Norman castle and cathedral. However, when Henry VIII established a naval dockyard in neighbouring Chatham, this became the more significant town. In Henry's time, the Medway at nearby Gillingham was already being used as a safe anchorage for the royal fleet, so this part of the river was an ideal location to start shipbuilding. Henry's original dockyard grew and grew over the centuries, as did its additional fortifications. The Napoleonic-era Fort Amherst, a complex of ditches, walls, tunnels and gun emplacements, was built to defend the dockyard from attack from the landward side, and later the army's Royal School of Military Engineering was established nearby. Visible from Ranscombe Farm Reserve is the Naval War Memorial on the Great Lines, one of three identical memorials established after the

First World War at Chatham, Portsmouth and Plymouth. Those familiar with Graham Swift's Booker Prize-winning book, *Last Orders*, or with the subsequent film, may remember the scene when the story's central characters visit this memorial on their way from South-east London to Margate. The Naval Dockyard at Chatham closed in 1984, and has mostly been redeveloped, with its 18th Century heart now forming Chatham Historic Dockyard, effectively a museum of naval shipbuilding. Former naval buildings now accommodate parts of the University of Kent and the University of Greenwich, and some of the buildings in the Historic Dockyard can be distantly glimpsed from parts of Ranscombe.

Roman Watling Street and the River Medway meant that Rochester, Chatham and the other Medway Towns always had great transport links. Boat travel was particularly important until eventually superseded by rail and then, more recently, road. The cement industry that became established along the tidal Medway after around 1850 was particularly dependent on the river not just for transporting the finished product, but also for providing the clay which – along with chalk – was a vital raw ingredient. This clay was excavated from the saltmarshes of the outer Medway Estuary downstream of Chatham, extracting the fine, compacted material that the river had originally cut from the Weald, carried downstream and deposited on the edge of the sea. So much material was extracted for cement-making that the whole shape of the Medway estuary changed, as islands and mud-banks were dug out and carried away. Inland, chalk was quarried from the hills of the Downs, leaving a series of massive pits along the scarp above the river. The river itself became lined with cement works, with a peak in the early 20th Century: in the 1910s, there were nine separate cement factories sitting along the Medway upstream of Rochester. All are now closed: the remains of the older works and their docksides have been retaken by nature, while later sites have been redeveloped for business or housing. The associated quarries have been filled in or built in, or in many cases just fenced around and left. The vanished industry still leaves its marks in local land-ownership, however: much of the local North Downs – including what is now Ranscombe Farm Reserve – was acquired by cement companies with a view to future quarrying, and the successors of the now-vanished companies still own very extensive areas.

The importance of the river for transport was to decline as rail travel took off in the early to mid-19th Century. The development of the railways in Britain is now so much a part of our national story that it looks to us like a steady process of continued progress and advance, but it doesn't take a lot of searching online before you realise how fragmented and disorganised the early railway systems were. Companies competed for access to certain towns and routes, which explains why there are two separate, mainline routes from Central London to the Medway Towns, one of which, the line that runs from London Victoria to Chatham, now forms the southern boundary of Ranscombe Farm Reserve. In the 19th Century, and well into the 20th, steam ships were still an effective and efficient way of travelling out from London to towns on the North Kent coast. But places like Rochester and Chatham, and especially ports like Dover, became obvious targets for establishment of the new and faster railway services. By 1849, the South Eastern Railway Company had opened a line from London to the Medway Towns, its route running close to the Thames through Dartford and Gravesend. So when the separate London, Chatham and Dover Railway Company got permission in 1853 to link the Medway Towns with Faversham, further east along the North Kent Coast, it hoped to make use of South Eastern's existing track from London to run services all the way to Faversham from the capital, with the aim of eventually going further eastwards to Canterbury and then Dover. But South Eastern Railway said no. This meant that London, Chatham and Dover had to build its own parallel line from London to the Medway Towns. This line opened in 1860 and ran south of the other rail route, riding up the dip slope of the North Downs; above the little village of Cuxton it ran along the side of the valley via a series of cuttings and embankments, before turning to the north-east to follow the River Medway up to Strood and Rochester. In following this route, the railway chopped up woods and fields and – as we'll see – sparked other landscape changes; but an incidental result of this 19th Century lack of co-operation was that there was a solid boundary to limit the expansion of Cuxton in the 1960s. A lot of valuable wildlife habitat was undoubtedly lost to the new housing built at Cuxton in the 1960s (and which included the house where I grew up), but without the restriction of the railway, a much bigger area might have been lost and Ranscombe Farm Reserve might never have been born.

To the east of Ranscombe Farm, it was a major road which held back the tide of built development and would eventually define the whole eastern boundary of the nature reserve. The expansion of housing on the fringes of the town of Strood started in the 1930s and continued right through to the 1970s, the whole town pushing westwards towards Ranscombe and swallowing up woods and fields as it went. However, it was to be halted by the barrier formed by the M2 motorway, which had been constructed as a new trunk road in the first half of the 1960s. The new motorway took traffic from the A2 on a big, looping bypass around the Medway Towns to provide a faster link between London and Dover. Diverging from the straight line of Roman Watling Street to the west of Strood, the new road swept southwards to cross the Medway on a high viaduct across the river upstream of Rochester. As with the railway, new housing would eventually

Small scabious *Scabiosa columbaria* and other plants on the margin of Longhoes Field. Photograph: R I Moyse/Plantlife.

be built up to the solid barrier of the road, but not beyond. Later, the Channel Tunnel Rail Link (or High Speed 1) would be built alongside the M2, and it is this road-and-rail corridor which now forms the eastern boundary of Ranscombe Farm Reserve. Like the walls around medieval Rochester, or the ditches and ramparts that protected Chatham's naval dockyard, these major transport routes acted as a shield for the land that now forms Ranscombe Farm Reserve. Tucked into its corner, the reserve butts right up against the Medway Towns but remains relatively unaffected by them.

This, then, is the context for Ranscombe Farm Reserve. It has sat on the border of major cultural, military and industrial upheavals for the best part of two millennia. The landscape of Ranscombe has been shaped, changed, cut and battered by the changes all around it, so that it is not just another bit of countryside on the margins of a big city, but as much a part of the Medway Towns and their history as Rochester Castle or Chatham Dockyard.

We've got the setting, so now let me take you on a tour to get the lie of the land. Our start is the main entrance to the site, which is more-or-less in the lower right-hand corner, close to where the Channel Tunnel Rail Link crosses the London Victoria mainline. Walk in the main entrance and you are heading roughly northwards up the steep slope on the west side of the Medway Gap, with the river itself lying some 50m (165ft) below you. To your left, as you go up the slope, is a long, narrow field called Longhoes which, just on its own, could probably be considered one of the country's most important sites for rare plants; the soil here is thin and dry, over the chalk that has been exposed on the valley sides. Ahead, the path continues up through woodland until you come out on a wide plateau with more woodland on your right – this is Merrals Shaw – and a broad arable field on your left. The level ground here has formed over clay-with-flints, with the flints so dense that Ranscombe's tenant farmer calls the big arable field 'Brighton Beach', because it reminds him so much of that resort's shingle shoreline.

Where you are now standing is part of the original Ranscombe Farm, which occupied all the flat farmland around you. Ahead, you can see the cluster of buildings that were once the farm's hub. Today the farmhouse, barn, oast house and other farm buildings have been converted into private homes and the land around them is not accessible to visitors. However, they were working buildings until the early 21st Century, some 500 years or more after the farmhouse is thought to have been built. As a quick aside for music fans, one of the farm buildings was the original home of Ranscombe Studios, who still operate under that name and have recorded many local artists, including the James Taylor Quartet and the influential and singular Billy Childish. On the horizon, beyond the buildings, you can see Cobham Woods, which make up almost half the area of the whole reserve.

The road leading to the houses is private, but you can take a left turn and follow the line of the North Downs Way, the long-distance footpath which passes through Ranscombe Farm Reserve on its 156-mile journey from Farnham in Surrey to Dover on the East Kent coast. The path takes you through a thin strip of woodland, and as you pop out the far side, you find yourself with a broad view of the Medway below and the North Downs stretching away towards the south-west. Keep going and you drop down the side of a small valley; across the valley is a narrow, wooded ridge called Mill Hill, and hidden behind and below the ridge is the railway and the village of Cuxton. Also tucked away out of view behind the ridge, and more than half-a-mile from where you are standing, are the extensive, south-facing, grassy slopes of Brockles Field, which form the lower left-hand corner of the reserve.

These ripples in the landscape – the shallow valley in front of you and Cuxton's Bush Valley beyond – were formed by the same wearing effects of weather and water that took the top of the Weald-Artois anticline. On the valley sides, thin, pale soils have formed directly over the exposed chalk. The shallow valley has various names: previously it was part of the nearby (now vanished) Court Lodge Farm, and was divided up into fields with names including Knock, Little Bottoms and Nether Great Bottoms. Ranscombe Farm's tenant farmer calls it Cogger's Valley after a previous occupier and local farmer, but we just know it as the Southern Valley. At the top end of the valley, just beyond a small block of woodland and at the western edge of the reserve, is Kitchen Field: sometimes known as the Cuxton Weedy Field, it was Kitchen Field and the area around it which first made Ranscombe Farm famous as a location for rare plants.

For now, though, we are going to turn our backs on the Southern Valley, and take a path which branches off in a roughly northerly direction. This will take us up and out of the valley and back onto the flat farmed fields and their clay soils. We are now on the far side of the former farm buildings, and, if you can manage to see over the hedges and trees, you can take in the view eastwards across the Medway Towns and may spot the Naval War Memorial in Chatham on the skyline. From here, the path dog-legs to the left and takes us to the edge of the large block of continuous woodland that occupies the northern part of Ranscombe Farm Reserve. This is part of the even larger Cobham Woods, which extend well beyond the boundaries of the reserve. The part of the woodland within the reserve has a variety of names on maps, often labelled as Great Wood and with various sections named as Birch Wood, Broad Oak Wood, Clay Pond Wood, and Stogarts and Drapers Wood, though the origins of these names are lost in history and the boundaries between the various sections are unclear on the ground. Follow the footpath westwards along the southern edge of wood and you will find it drops gently down to meet the top of the Southern Valley at Kitchen Field. Alternatively, take one of the paths into the woodland, and you will find the ground rising to the highest point of the reserve: here, in the north-west corner of the site, the distance from where you started is a mile-and-a-quarter as the crow flies and you are 80m (260ft) higher. The geology has changed too, and you are on one of those deposits of sands and gravels that are dotted across the ridge of the North Downs.

From this high point, if you head eastwards through the woods you'll eventually meet the High Speed railway to the Channel Tunnel. Follow the footpath over the railway and under the M2 motorway and you'll find yourself among the houses and flats of Strood and just a few minutes' walk from Roman Watling Street in the shape of the modern A2. Alternatively, you can head to the west on the same path and be taken out of Ranscombe

Kitchen Field - also known as the Cuxton Weedy Field - the field that first made Ranscombe famous. Photograph: R I Moyse/Plantlife.

Farm Reserve and into Cobham Park, now managed by the National Trust. The park is a beautiful area of wood-pasture which was originally laid out in the late 18th Century by the prominent landscape designer Humphry Repton; his employer was the 4th Earl of Darnley who had recently inherited the house which was originally at the heart of the park, Cobham Hall, and its surrounding estate. Wood-pasture consists of trees scattered more-or-less densely in grazed pasture-land: as an economic use of the land, it would have once been more frequent on wooded commons where locals had a right to run animals, or in the parks and forests owned and managed as hunting grounds by royalty and the aristocracy. At Cobham, the wood-pasture was deliberately created as 'pleasure grounds' for the owners and occupiers of the Hall, and laid out to provide an idealised facsimile of a wild landscape, with landmark trees and carefully created views. There was even a decorative 'model dairy' (now restored as a holiday let by the Landmark Trust) where the Hall's ladies would play at being dairymaids.

Cobham Hall, today a private school, has a centuries-old history, a history which swept up and eventually spat out the land that became Ranscombe Farm Reserve. The oldest parts of the current house are Tudor, and the house and estate have been in the ownership of various families over the centuries. From the early 13th Century, the estate was owned by successive generations of the Cobham family until the 11th Lord Cobham got caught up in a plot against King James I. Then there were the Dukes of Richmond, until the 6th Duke died and the estate, with huge debts, was inherited by his sister, Katherine Stewart, who went on to marry the politician and diplomat Sir Joseph Williamson. In the early 1700s, after the death of Sir Joseph Williamson, the hall and estate were purchased by John Bligh, who was later granted the hereditary title of Earl of Darnley, and the land remained in this family until the 20th Century, when first the outlying estate and later the hall and park were sold.

Students of cricket history will remember that it was Ivo Bligh, later the 8th Earl of Darnley, who captained the England team to victory in Australia in the first 'Ashes' test series. The previous series of games between the two countries had been held at The Oval cricket ground in South London, and the England team was so badly beaten that a mock obituary was published, stating that the body of English Cricket would be cremated and sent to Australia. Accordingly, when England then won in Australia, Bligh was presented with a small urn, containing those supposed ashes, to take back to England. This urn was displayed at Cobham Hall until Ivo Bligh's death and is now on show in the MCC museum at Lord's Cricket Ground in London. It is said that Ivo Bligh's cricketing expertise may owe a lot to local bricklayer and international cricketer, George Bennett, who had played in the first cricket tour to Australia in 1861-2 and is believed to have played the game with the young Bligh.

The land that became Ranscombe Farm Reserve seems to have been part of the Cobham Hall estate for centuries. This was certainly the case in the first half of 19th Century, when tithe schedule shows that all of today's Ranscombe Farm Reserve was owned by the Earl of Darnley. The schedule was drawn up as part of changes to the ancient tithe system – essentially a tax on the value of land – and shows that most of the open, farmed fields that now make up the reserve fell within Ranscombe Farm and Court Lodge Farm, though both farms were at that time tenanted by the same person, William Pye. The Pye family had a long history of farming in the area, and their names still live on in the roads that make up much of Cuxton's 1960s housing estate, including William Road, James Road and Hayley Close.

The history of Cobham Hall also ties up with my own personal history. Sir Joseph Williamson,, who became owner and manager of the estate through marriage, left £5,000 in his will "for and towards the Building, perfecting, carrying on and perpetuall maintaining of a free Schoole at Rochester and of a schoole Master or School Masters for the Instruction and Education of such Youth there who were or shall be Sons of Freemen towards the Mathematicks, and all other things which may fitt and Encourage them for the Sea Service or Arts and callings leading or Relating thereto". Sir Joseph Williamson's Mathematical School is still going today as an academy school; I was there between 1974 and 1981, when it was a county grammar school under a selective education system that set your choice of schools based on your results in the 11+ exam. At the time I arrived, the school had not long moved from its original site in Rochester Town Centre (now a car park by the remains of the old city wall).

What was never mentioned in my day, but is stated – to their great credit – on the school's current website, is that Sir Joe was a slaver, and in a big way. He was an investor and administrator in the Royal African Company, which for decades had a monopoly on the English slave trade and is thought to have shipped more enslaved Africans to the Americas than any other company in the history of the transatlantic slave trade. A number of websites quote an average figure of 5,000 people kidnapped and enslaved each a year over the period 1680-86, and the company lasted for 60 years before its demise in 1731. This appalling, murderous enterprise may have had an impact on at least some of the land in Ranscombe Farm Reserve: an 1877 article in the journal of the Kent Archaeological Society, *Archaeologia Cantiana*, reporting on Williamson's will, notes that "his Cobham estates extended to the town of Gravesend on the one side and to the city of Rochester on the other" and this must have included part, probably all, of the current reserve. As the estate was in debt at the time of

Williamson's marriage to Katherine Stewart, it seems reasonable to suppose that – just possibly – Ranscombe Farm Reserve might look different today if it hadn't been for the profits of slavery. I am no historian, and I could, of course, be entirely wrong, and this is certainly not intended as a guilt-trip for myself or anybody else. However, we have to know how we got to where we are now or we will not be able to move forward effectively; this makes it essential to acknowledge that the world today is, at least in part, shaped by the terrible wrongs of the past.

There is certainly one direct line that can be drawn from the 17th Century to modern-day Ranscombe Farm Reserve, and it is one that matters hugely to our story because it concerns botany and botanists. For, at the very same time that Sir Joseph Williamson was master of the Cobham estate, early botanists were visiting the area and making the first written record of meadow clary *Salvia pratensis* growing in the wild in Britain. The discovery of this blue-flowered wild sage as a British wildflower can be seen to mark the start of Ranscombe Farm's journey towards being recognised as a site of exceptional interest for wild plants, and it played a part in the establishment of the site as a nature reserve more than 300 years later.

It was in the 1600s that plant-hunting started to get big. In 1597, John Gerard had published his famous Herball, describing both wild and cultivated plants and their medicinal uses. This was revised, updated, and corrected in 1633 by the London-based apothecary (effectively, an early pharmacist), Thomas Johnson, who added notes on where he had found plants during his own excursions. As noted by Richard Mabey in an article in the journal *British Wildlife*, Johnson used to organise plant-hunting field-trips for apprentices from the Society of Apothecaries, this being one of the Livery Companies set up in the City of London to control and manage the practice of certain trades – in this case, pharmacy and medicine.

In this period, road travel was slow and difficult, and boats offered the best option for covering any distance. As a result, North Kent became a handy target area for any early botanists working out of London: you could go some way on foot, or take a boat to give you a starting point further out from the city. In 1629, Johnson and his associates made two trips to Kent, which are described in his *Iter Plantarum*, one of which started by taking a boat to Gravesend, from which they proceeded, largely on foot, to Rochester, Gillingham and eventually Queenborough on the Isle of Sheppey (a distance of well over 20 miles) before returning in a loop via the

Isle of Grain. This took several days, and the description of the trip, which includes extensive lists of the plants found on each leg of the journey, can be found in translation (from the original Latin) online. It was this and other well-documented excursions, together with his extensive knowledge, that have led to Thomas Johnson being called the father of British field botany.

The 1629 trips did not take in the woods and fields around Ranscombe, but the closeness of this area to Watling Street and the City of Rochester (where food and accommodation were readily found) meant that it fell within easy reach of the early botanical explorers. In his 1633 edition of Gerard's Herball, Johnson notes that he has, himself, found yellow archangel *Lamiastrum galeobdolon* growing in "the woods belonging to the Lord Cobham in Kent". So Johnson was in the area and might possibly have set foot in what was to become Ranscombe Farm Reserve. However, we have to wait until 1696 to get the first definite evidence of visiting botanists, when the finding of meadow clary was reported by Leonard Plukenet in his grandly titled *Almagestum Botanicum sive Phytographiae Plukenetianae Onomasticon*. Plukenet as one of the most prominent botanists of his day: he was in charge of the royal gardens at Hampton Court during the reign of William and Mary, and his collections of pressed plant specimens and pressed (yes, pressed) insects are amongst the founding collections of the Natural History Museum in London.

Thanks to details given by the Kent botanist, Geoffrey Kitchener, in the Kent Rare Plant Register Draft Species Accounts, I was able to track down Plukenet's *Almagestum Botanicum* online and better able to understand what it says. The full title of the book seems to roughly translate as "Great Botanical Treatise or Plukenet's Alphabetical list of Plant Descriptions", and it consists of around 400 pages of densely-written Latin, complete with abbreviations, that has the potential to baffle many web-based translators (and which is well beyond my two years of Latin at Sir Joseph Williamson's Mathematical School). Not only that, but the book was published long before Linnaeus came up with the system of giving every species a standard, two-word Latin name. So meadow clary which botanists today know as *Salvia pratensis*. was called by Plukenet *Horminum pratense foliis serratis* – something which made looking it up somewhat more difficult than I expected.

But, having finally found the original reference, and more-or-less cracked the translation, it does seem clear that Plukenet recognises meadow clary as a plant already known to and grown by herbalists – it was used to treat eye disorders – and that he also knows that it is not the same plant as the similar, but much more common, wild clary *Salvia verbenaca*. He notes that it has been "recently discovered at least once (if the narrators be credited) in the Kentish Park of Cobham". Meadow clary is a good-sized

The plant that started it all meadow clary *Salvia pratensis*, Plukenet's *Horminum pratense foliis serratis*, still growing on the edge of Cobham Woods. Photograph: R I Moyse/Plantlife.

plant, with large, crinkly leaves and able to form substantial clumps. In early summer, the basal rosette of leaves sends up flower-stems that are a couple of feet (60cm) or more in height and which bear dazzlingly-blue, two-lipped flowers. It was subsequently found growing elsewhere in the wild in Britain, but it has always been rare and its discovery in Cobham Park seems to have been a lure drawing botanists to the area ever since. Visit Ranscombe Farm Reserve today, and you can still find meadow clary growing along the south-facing slope at the very edge of Cobham Woods and easily visible from the footpath: knowing what we do about the past and present distribution of the plant in Kent, we can be certain this place – or very close by – was the scene of the first discovery.

It may have been a trip to hunt for meadow clary that led a later botanist to become the first person to find rough mallow *Malva setigera* growing wild in Britain. Rough mallow (sometimes called rough marsh-mallow or hairy mallow) is a low-growing annual or biennial plant of disturbed ground. Trailing stems, armed with stiff, bristly hairs, spread out from a central rosette of rounded leaves, and bear attractive, pale-pink flowers. The first recorded occurrence of the plant in the wild in Britain was noted by the

Reverend Jelinger Symons in his *Synopsis Plantarum insulis Britannicis indigenarum* ("Synopsis of plants native to the British Isles"): "*In arvo prope Cobham in Cantia anno 1792 invenit J. Rayer*", which translates as "found by J. Rayer in a ploughed field near Cobham in Kent in the year 1792". Like meadow clary, rough mallow is and always was a rare plant in this country, and, also like meadow clary, it is still found today at Ranscombe Farm – and still in some of the ploughed fields – but not elsewhere around Cobham.

The presence of these two notable species continued to draw visitors to the Ranscombe area during the 19th and 20th Centuries, and, indeed, it still brings plant-lovers to Ranscombe Farm Reserve today. Most prominent amongst those visiting during the 19th Century was probably the Reverend Professor John Stevens Henslow. Henslow had been born at Rochester in 1796, and by 1825 was professor of botany at Cambridge University. In 1828, the young Charles Darwin began studying at Cambridge, where Henslow became his tutor and lifelong friend. It was Henslow who recommended Darwin for the post of naturalist on the famous voyage of HMS Beagle which kick-started Darwin's thinking on evolution; Henslow himself had been the first choice for the trip but had turned down the offer. We can be certain that Henslow came to Ranscombe Farm, because just a year before his fateful meeting with Darwin, he collected and pressed a specimen of rough mallow. The specimen is labelled as "*Althaea hirsuta* hispid Marsh-mallow nr Cobham, Kent July 1827" and is in the keeping of the South London Botanical Institute but can be viewed online thanks to the herbaria@home project (a project which provides access to high quality images of a massive number of pressed and dried plant specimens, and is made possible by an army of citizen scientists and supported by the Botanical Collections Managers Group and the BSBI).

The early world of natural history seems to have been mainly inhabited by scholars and those men (and it was mainly men) wealthy enough to have plenty of leisure time, which often meant members of the clergy. This learned atmosphere is no doubt why some publications were still appearing in Latin as late as the end of the 1700s, when the Reverend Jelinger Symons published his *Synopsis Plantarum*. But the rise of the middle class in the mid-19th Century led to a great many more people having the time and money to pursue interests like the study of natural history. New societies were created and new publications launched. The year 1836 saw the founding of the Botanical Society of London, which is still going strong today as the Botanical Society of Britain and Ireland (BSBI): just three years later, the society's published Proceedings included a report on a "Botanical Excursion to Cobham and Cuxton, Kent" made in 1838 by an unnamed group of the society's members. The trip started in Woolwich, in

The herbarium sheet with the specimen of rough mallow *Malva setigera* (under its old name of hispid marsh-mallow *Althaea hirsuta*) collected by the Rev. Prof. John Stevens Henslow and the Rev. William Strong Hore in 1827. The original sheet is in the herbarium of the South London Botanical Institute, which has kindly granted permission for the use of this image.

South-east London, and appears to have been made on foot, as there are remarks on plants found all along the route – a journey of more than twenty miles each way. Those concerned made a second, later trip of the same distance.

Interestingly, on the first trip, this particular group of botanists appear to have found meadow clary growing amongst newly planted conifers in an early 19th Century plantation on the southern edge of what is now Ranscombe Farm Reserve – a place where the plant no longer occurs and some distance from where it was originally discovered. On their second trip, the excursionists found their way to the southern edge of Cobham Woods, finding both meadow clary and rough mallow "in abundance" in that part of the site where the two species can still be found growing close together. Not long after this trip, in the first volume of The Phytologist (a botanical journal which ran for some years in the mid-19th Century) is a report of a similar botanical excursion made in 1841. A Mr Alexander Irvine and his friend Mr William Pamplin (later to be editor and publisher respectively of The Phytologist) took a steam ship for the 2½-hour trip from London Bridge to Gravesend, before heading inland on foot to Cobham and Cuxton – giving themselves a round trip of just 16 miles – and again finding meadow clary and rough mallow in the fields below Cobham Woods. Irvine returned to Ranscombe in 1863, when he collected specimens of meadow clary which became part of the collections of the Manchester-based Charles Bailey and the very prominent botanist George Claridge Druce, who was made a Fellow of the Royal Society and at one time served as Mayor of Oxford; these specimens are now in the herbarium of the Manchester Museum, part of the University of Manchester.

Clearly, the patch of land between Cobham and Cuxton which we know today as Ranscombe Farm Reserve was already famous as a location for finding rare plants by the middle of the 19th Century. Among the 166,000 specimens digitised by the herbaria@home project, there are twenty-five pressed specimens of rough mallow which were collected at Ranscombe Farm in the 110 years following Henslow's visit in 1827. In most cases, it looks as if the visiting botanists came specifically to find rough mallow and meadow clary, though some also went away with samples of other notable plants, like broad-leaved cudweed *Filago pyramidata* and ground-pine *Ajuga chamaepitys*, that I'll say more about in due course.

Few of these Victorian and Edwardian botanists were academics, as it is really this period which marks the rise of the amateur naturalist as the driving force behind the study of British wildlife, something that is still the case today. William Pamplin was a bookseller by trade; Charles Bailey worked for an international shipping company; Druce was a pharmacist. A later visitor was J. E. (Ted) Lousley, whose career was in finance as a

Rough mallow growing in Kitchen Field, the same spot where Lousley found it in 1933. Photograph: R I Moyse/Plantlife.

banker and at the London Stock Exchange, but who became another exceptional botanist and left on his death in 1976 a collection of just short of 25,000 pressed and labelled plant specimens. In 1950, Lousley's book, *Wildflowers of Chalk and Limestone*, was published as part of the hugely important and influential New Naturalist series (which is still going and now with more than a hundred titles). In the book, Lousley writes of a place on the west side of the River Medway where rough mallow (he uses the older name of hairy mallow) "has led a somewhat chequered existence for at least a century and a half", and notes that alongside it grow ground-pine and broad-leaved cudweed, with "our oldest locality for meadow clary" being not far away from this spot. With this selection of plants, he can only be talking about Kitchen Field at the top of Ranscombe Farm Reserve's Southern Valley. Thanks to herbaria@home, we can see that Lousley himself came to the site in 1933, when he followed the fashion of the previous two centuries by collecting specimens of rough mallow (which he labels as "abundant in cult. [i.e. cultivated] ground along edge of Great Wood") and meadow clary.

Lousley and his predecessors are part of the history of Ranscombe Farm Reserve and its wild plants. It is a history that is influenced by the great movement of tectonic plates that pushed up the rocks so that they weathered in the way they did to form the shape of the slopes and valleys. It takes in the Roman invasion that led to the establishment of Rochester and the Medway Towns, and of Watling Street as an important transport corridor. Something is owed to the changing fortunes and families of Cobham Hall and its surrounding estates, as well as to the uncoordinated nature of the 19th Century railway boom. And a great deal rests on the site's accessibility to London's apothecaries and to the amateur botanists that learnt from them and – literally – followed in their footsteps. There is more, of course: there is an astonishing range of rare plants that have made a home in Ranscombe Farm Reserve, and they all have their own histories and back-stories. But that is for a later chapter.

Chapter 2 - The making of a nature reserve

My own first encounter with rough mallow was, I suppose, sometime in the late 1970s or early 1980s. My mother had been out for a walk up to Kitchen Field and said she had seen a plant that she didn't recognise and that it had pink, five-petalled flowers and rounded leaves. Going back with my copy of Fitter, Fitter and Blamey, we were able to properly identify the mystery plant. This was some decades before Ranscombe became a nature reserve, when it was privately-owned, though criss-crossed with public footpaths which made it nearly as accessible then as it is today. I can still remember not just rough mallow but a lot of the other plants I found and where I found them. The Southern Valley at Ranscombe Farm was an easy walk from our house: down the road, left into Whiteleaves Rise, through the arch under the railway and there you were. A well-trodden track (later ploughed out but since reinstated) ran up the bottom of the valley, and it was on in field edges alongside the track that I first encountered Venus's-looking-glass *Legousia hybrida* with its rather clumsy-looking, wavy-edged leaves offset by its small, elegant, purple flowers. Nearby, and up amongst young woodland, and in fairly deep shade, I once came across a few man orchids *Orchis anthropophora* in an unlikely location for what is generally thought a plant of open grassland. Best of all was walking along the edge of Brockles Field: this field is a wildflower meadow today, but back in the 1970s it was an arable field, and on one margin I found a small few plants of ground-pine. I knew even back then what a rarity this little, furry, grey-green thing was. No doubt others had been long aware of its growing there, but the thrill of making your own discovery of something so scarce – and on your own patch – never leaves you.

Venus's-looking-glass.
Photograph: R I Moyse/Plantlife.

This interest in the natural world was something that infected all four of our family, but it was not inevitable. My parents were really townies, my mother having grown up in the South-east London suburb of Beckenham in a two-up, two-down terraced house with no bathroom, and an outside toilet which was lit only by a small paraffin lamp. Her parents lived there till their deaths in the 1970s, and I still find the smell of paraffin takes me back to our regular weekend visits. My father was from slightly more up-market Bexley. My parents had no great interest in natural history when they came to Cuxton (and, indeed, my mother at first found the emptiness of the countryside slightly unnerving); they were just a couple of working people who were not especially (formally) educated but who were clever and curious and interested in their surroundings. They also had a dog, which needed walking, and two children, who needed entertaining and exercising, so we used to go 'up the woods' quite a lot, and pretty much from the start.

On clearing out the house after my father passed away, we found a small black-and-white photo of myself and my sister sitting on a blanket in a patch of grass. Behind us are the concrete posts of a wire fence and beyond that some low trees. On the back of the photo, and in my mother's handwriting, it gives the year: 67. Given that I'm in a t-shirt, it must be summer, so that means that I would have been almost exactly four years of age when the picture was taken, and my sister was just a year old. The location is instantly recognisable because the little patch of grassland is still there, as is the wire fence and its concrete posts: it is on the southern edge of the reserve, on the chalky ridge that separates Ranscombe Farm Reserve's Southern Valley from the deeper valley in which the village of Cuxton sits. Behind and below the wire fence, in a deep cutting, runs the main railway line from London Victoria. Today, the grassland where we picnicked is the part of Ranscombe Farm Reserve known as Mill Hill.

It was finding ourselves living in a place with easy access to the countryside and with woods, fields and riverbanks to explore, that got us all learning about wildlife together. I know that I had already developed an interest in the natural world while I was at junior school, because I remember on at least one occasion saying that I wanted to be a naturalist when I grew up, as if it were a job in itself, like being a train driver. I had been told that the man who presented the children's programme *Fabulous Animals* on the television was a naturalist, and I liked the programme and its stories about animals both real and imaginary, and thought that working in that sort of world would be great. That presenter was David Attenborough – as I'm sure some of you will remember – and I still have the little paperback book, published in 1975, that accompanied the series, with its tales of mermaids and manticores and seals and Komodo dragons.

None of our family would have got anywhere with wildlife if we hadn't been able to learn from other people. There were of course field guides that could be bought, though fewer than today, and we would still use things like

The author (in stripy t-shirt) and his sister in the grassland at Mill Hill in 1967. A visitor today would see a more heavily wooded scene.

the tiny and popular (and now collectable) Observers books. There were also nature reserves that could be visited, though again fewer than today, and often you needed to get a permit from the warden first, which might mean knocking on a door or writing off in advance. But the best way to find out about what books to get and what places to visit was to take advice from someone more experienced. This is where the groundwork put in by those Victorian corresponding societies and natural history clubs paid off – and continues to pay off – for all of us, as it meant that there were groups to join and which were happy to welcome new members. Locally, the Rochester and District Natural History Society had been founded in 1877, after which it ran for another 125 years, publishing *Natural History in the Rochester Area* (which includes a guided walk round Ranscombe Farm) on its centenary in 1977. My parents later joined the Rochester 'Nats', but back in the 1970s, we were members of another local club, the sadly now also defunct North Kent Wildlife Preservation Society. It was going out on field meetings with this group that started to turn me away from looking mainly at birds and more towards plants, particularly under the influence of some of the mighty botanical brains there, including the late Mervyn Brown who will be remembered by lots of people in Kent's natural history community.

I was also lucky that the leader of the local Cub Scouts was the bearded and affable Ken Law, who had lived locally for (I think) his whole life. I don't remember much of my time as a Cub, but I do remember that one summer evening the whole troop went on a walk up to Ranscombe Farm. We followed the route of the North Downs Way up the slope of the Southern Valley, where Ken pointed out and gave the names of the different trees and shrubs. The woodland alongside the path here is a narrow strip which follows the old boundary between the original Court Farm and Ranscombe Farm holdings. It still has a mix of field maple *Acer campestre*, wayfaring-tree *Viburnum lantana*, wild privet *Ligustrum vulgare*, pedunculate oak *Quercus robur*, blackthorn *Prunus spinosa*, dogwood *Cornus sanguinea* and common whitebeam *Sorbus aria*: the kind of rich scrub that can often develop on chalky soils. I recall that Ken referred to common whitebeam as 'silver hornbeam', and I hope I'm right in remembering this, as the name is not in Geoffrey Grigson's *The Englishman's Flora* (which gives so many of the folk names for wild plants) but seems spot-on for a tree with such hard, white wood and whose leaves are so strongly silvered by their coat of white hairs. Ken has a genuine passion for wildlife and when I returned to work at Ranscombe Farm some 37 years after leaving the Cubs, he was one of the very active team of volunteers that had been – and still are – crucially important to the management of the reserve.

One final Ranscombe Farm reminiscence from my schooldays. I was incredibly lucky that my secondary school – the one founded by Sir Joseph Williamson – offered botany as an A-level option, which may well have been down to one of the biology staff, Bob Foulds, having a botanical backgrounded. As a result, not only was I encouraged in my interest, but I was also able to study botany at A-level and to follow my ambition to study botany beyond school. During one of the breaks between terms, our little A-level group of botanists were set a project. Exactly, what we were supposed to do, I cannot remember, but I do know that as part of it, I went up to the patch of grassland at Mill Hill where we used to picnic, with my plant book, and set about compiling a list of the species I could find. This something I wish I still had, as I am sure that I found dropwort *Filipendula vulgaris*,

Dropwort *Filipendula vulgaris* - a plant no longer found at Ranscombe.

which is like a small, downland version of the wetland plant meadowsweet *Filipendula ulmaria*, and has a similar froth of white flowers on top. Dropwort does not grow at Ranscombe today, which makes me wonder when it might have disappeared and why, and what else may have vanished in with it.

A-levels gave way to university, and I went on to study plant biology at Bangor in beautiful North Wales. In early 1986, I was back in the Medway Towns, working for the then Gillingham Borough Council, which meant I was close to Ranscombe Farm again. I was able to witness changes like the gradual loss of the grassland at Mill Hill where I had picnicked as a child, as trees and shrubs moved in. I also saw the massive, immediate change that resulted from the Great Storm of 1987. This felled large parts of the woodland on the south-facing slopes at Mill Hill, including some huge, old beech *Fagus sylvatica* trees that must have been there for at least a century, though in doing so it also opened up bare ground for plants like wild columbine *Aquilegia vulgaris* and horseshoe vetch *Hippocrepis comosa*, neither of which I had found there previously.

The world in which I kicked off my own personal journey as a naturalist is disappearing. Local natural history groups like the North Kent Wildlife Preservation Society and the Rochester 'Nats' have gone, with those sort of neighbourly clubs replaced by a wide range of friendly and supportive groups and individuals on social media. Real-world, face-to-face clubs and societies still abound, but are increasingly specialised: there are county-based bat groups, reptile and amphibian groups, mammal groups, botanical recording groups and more, reflecting the increasing specialisation and depth of knowledge shown by today's amateur naturalists. We get the organisations, both amateur and professional, that we need for the times – in many ways, the loss of the old natural history societies shows that they did their jobs in creating a wider, more democratic, more in-depth world of natural history study. Continuing to drive this process forward today are bodies like Birding for All, Gay Outdoor Club, Black2Nature and Flock Together, all of which are tackling the barriers that still face many people who want to get out into nature – barriers that I've never had to face as a 59-year-old, straight, white man and not living with any disabilities or significant health issues.

If we get the natural history clubs that we need, then we also get the conservation organisations that we need, and it was while I was working for Gillingham Borough Council that I got to witness the start of a new plant conservation charity called Plantlife. In 1988, a conference titled 'Is nature conservation working for plants?' was held at the Zoological Society of London, and I had managed to persuade my manager at Gillingham Borough Council that my role there as a countryside ranger meant that

Columbine *Aquilegia vulgaris* growing at Mill Hill. The blue-flowered wild form is still common, though it is clear that there has been hybridisation with garden varieties, producing a range of colours and patterns.
Photograph: R I Moyse/Plantlife.

attending the conference was valid as part of my professional development. Towards the end of the day, when questions and debate were being directed through the chair, there was a minor revolt in the audience when one delegate stood up and quite vehemently made the point that plants were being failed by nature conservation and that there was a need for a new conservation body to stand up for plants in the same way that RSPB worked for birds. The chair was not impressed and did not think that this was an appropriate interjection, but clearly the delegate's view was shared by a number of people present, because they took up the cry and the cause, and the small gang of them went off to form Plantlife in 1989. Of course, the punchline here ought to have been that I was one of those people, but I was too naïve and reticent to really understand the value of what was happening, and it would be more than twenty years till I became part of the Plantlife team.

The new conservation charity found itself involved in Ranscombe Farm at an early stage. In the 1990s, it was running a project called Back from the Brink which was focused on a range of rare and threatened plants. Our old friend, rough mallow, was one of the plants on the list, and the work carried out at Ranscombe Farm included sending people out with garden forks to dig over ground in and around Kitchen Field – the site of the species' first discovery – to create bare ground and bring up buried seed from down in the soil. This proved spectacularly successful, with several hundred plants appearing, though this kind of intensity of management would be difficult to maintain in the long term. But rough mallow was not the only plant at Ranscombe Farm to get attention from the Back from the Brink project. Ground-pine and broad-leaved cudweed, both of which had been noted and collected at Ranscombe Farm by early visiting botanists (and mentioned by Lousley in his New Naturalist volume) were also targeted.

In fact, tiny Kitchen Field (it is not quite 5 hectares, less than 12 acres, in extent, meaning you can potentially stroll across it in just three or four minutes) became an increased focus of botanical attention during the later 20th Century, and not just for plants that had always been rare in Britain, like rough mallow and ground-pine. Starting with the agricultural revolution of the 19th Century, the weedy plants once commonly found in cornfields or other ploughed crop fields had been disappearing from the farmed countryside, and their decline had accelerated after the Second World War. Yet, astonishingly, many of these plants seemed to be hanging on in Kitchen Field – so much so that the field was made part of the Cobham Woods Site of Special Scientific Interest (SSSI) in 1984. The citation document for the SSSI is very much focused on the woodland which makes up the bulk of the site, but does note that "Also included in the site is an area of arable land which is notable for the number of uncommon annual plant species which occur" with name-checks for rough mallow, ground-pine, Venus's-looking-glass and blue pimpernel *Lysimachia foemina*.

By the late 1990s, Kitchen Field was becoming increasingly known not just as a spot with one or two rare plants, but as place of national significance for farmland flora, and it was firmly on Plantlife's radar as the 'Cuxton Weedy Field'. Because it had been designated as a SSSI, the government nature conservation agency, English Nature (now Natural England), was also working to conserve the field's wild plants by encouraging and supporting the farmer in reducing chemical inputs and especially herbicides. As a result of these combined efforts, by 2006, Peter Marren was able to call the Cuxton Weedy Field "possibly the best place for arable wildflowers in all England" in an article he wrote for The Independent newspaper.

Peter Marren was writing around the time that Ranscombe Farm Reserve opened as Plantlife's flagship site. Hailed as the "Medway miracle" by Professor David Bellamy at the official opening, the reserve took in Kitchen Field – the Cuxton Weedy Field – and wrapped it with nearly 240 hectares (not far short of 590 acres) of additional land. More than three centuries had passed since Plukenet had first put the area on the botanical map: can there be any other nature reserves with such a long-charted history from the very first recognition of their wildlife interest? Certainly, such places must be few, as so many of the sites of interest to early botanists are likely to have been lost to neglect or destruction or simple, slow change. So many accidents might have happened over 300 years, that the eventual arrival of Ranscombe Farm Reserve can look like no more than good fortune – Bellamy's "miracle". In some ways, this may be true, but this would seriously underplay the determination and drive of the people involved, from the 19th Century botanists on their 40-mile round-trips on foot, to the few dedicated people at the 1988 conference who created a new national charity from scratch, and the Back from the Brink workers who toiled with hand-tools on flinty Ranscombe Farm soil. Isaac Newton famously wrote, "If I have seen further it is by standing on the shoulders of Giants" to explain the huge strides he made in scientific understanding, and it often seems as if this idea applies as much to conserving nature as it does to understanding nature.

Though the image of a tower of people, each stood atop another, is perhaps inaccurate. If you look at just the period around the establishment of Ranscombe Farm as a nature reserve – say just five years either side of its official opening – then it starts to look more like a large and much-branched family tree. Acquiring around 600 acres of privately owned woodland and farmland, making it safe and secure, and bringing it into to some form of active nature conservation management is a complex, lengthy and costly process. Giving due credit to everyone who made the process possible could end up looking like one of the longer and more indulgent acceptance speeches at an awards ceremony.

Nonetheless, it is important to acknowledge that Medway Council played the key role, having already identified the site in its local development plan as potential new country park. It was Medway that was able to manage the planning process and to secure funding for most of the land purchase, this money coming from a government pot for creation of new greenspace in Thames Gateway. The Cobham Ashenbank Management Scheme (CAMS) also had an important early role, and had in fact been working in the wider area since 1997, when it was established as a multi-agency partnership which would work to compensate for the impact of the Channel Tunnel Rail Link on Cobham Park and Cobham Woods. Before CAMS came along, the woods and park were easily accessible to vehicles and suffered massive amounts of misuse: eventually, the project removed more than a hundred cars that had been dumped and burnt in the woods over the years.

There were also key individuals whose talents were critical to the successful birth of Ranscombe Farm Reserve. Medway Council was lucky enough to have the whip-smart and professional Martin Hall, who had more than enough knowledge of Kent's environment to be able to see the potential of the proposed new reserve to meet the needs of people and wildlife, and also had the nous and experience to successfully steer a path through the corridors of local and national politics. For Plantlife, Andy Byfield both knew the site's plants in detail and, probably better than anyone, could see and vividly articulate a grand vision for the new reserve. Andy is someone with an encyclopaedic knowledge of wild plants. In fact, more than that, he seemed to have the low-down on every species of British plant, and to have met all the really interesting species face-to-face; he can tell you where they grow, how they are faring, and what they need. He's also a superb field botanist, and I once saw him spot a single plant of chaffweed *Lysimachia minima* without bending over or barely even breaking stride – and this is a tiny scrap of a plant which bears just a few leaves less than half-a-centimetre in length and tiny flowers just a millimetre across and which I have gone on hands and knees to try and find.

Without political acumen and the ability to sell a big vision, no project of any scale is going to get off the ground. But when it comes to the crunch, you must have people with the skills, energy and perseverance to turn the vision into something real on the ground. In this respect, Medway Council's Chris Valdus was a calm, able and pragmatic presence when tackling snags and barriers in the reserve's early days. Plantlife itself was very lucky to have Joe Costley, who worked with solicitors to put together the complex agreements between Medway Council and Plantlife which would secure the site as a Plantlife nature reserve, and who oversaw management and fund-raising for the site over its first five or six years. For some time, this meant that Joe had to make the trip to Ranscombe Farm from his home in Salisbury on a more-or-less weekly basis (he told me he became very familiar with the Clacket Lane services on the M25), while at the same time retaining responsibility for twenty-two other Plantlife nature reserves. When I started work at Ranscombe Farm Reserve in 2011, Joe was able to drop this one site, but was still the only Plantlife staff member for a nature reserve network which stretched from the Isle of Man to Suffolk and from Cornwall to Cumbria. Although most of these reserves are and were let to partner organisations, particularly local Wildlife Trusts, the largest were still Joe's direct responsibility, from Joan's Hill Farm in Herefordshire, to Seaton Meadows in Rutland and Deep Dale in the Derbyshire peaks, and none had an on-the-ground warden. That these reserves are so good – and they are – tells of Joe's infinite capacity for taking pains, and a head for detail that meant that he was able to get Ranscombe off the ground at the same time.

In a typically self-effacing way, Joe himself is full of praise for Ranscombe Farm Reserve's first warden, Helen Morley, who was at the pointiest end of the job of creating a new nature reserve. Nature reserve wardens don't always get the credit they are due, but they are the people that have to directly wrestle with tools, machinery, the weather, the land, and, quite often, the will of other people in order to make the vision for a place into a reality. A compilation of the experiences of nature reserve wardens would make a fascinating and revealing book, and perhaps especially if it focused on female wardens, who find themselves working in the very male-dominated sphere of land management.

In Helen's case, this meant having to deal with being shouted at by a (male) local council officer over a problem with tree-felling licences, managing with the attitudes of (again male) forestry contractors, and even coping with comments from some volunteers. As Helen told me, "No one liked a woman operating a chainsaw and I was good at it and knew when a stool had been well cut or badly cut, and so I learnt a lot about contractors and life, and they learnt a lot about themselves from my perspective too! It was harder when the volunteers made snidey remarks about 'women drivers' as I had to take that from them, but sometimes it was really difficult." Despite having to work in part from the cab of a pick-up truck, having to store tools at home, and even being shot at by deer-poachers, Helen was able not just to put together a management plan for the reserve, but get the woodland into positive conservation management – together with securing grant aid from the Forestry Commission to cover part of the costs – and secure agri-environment funding to get the most important arable farmland managed specifically for rare wildflowers. She was also able to set up a new group of volunteers to help with management, about which she says, "they were on my side completely and so I called them my A team … At times I was somewhat overwhelmed with the responsibility of a large SSSI habitat deemed 'in decline' (or whatever the disparaging phrase was) pretty much on my own and with no budget, so they were a great encouragement as well as enabling me to achieve the ambitious management tasks we had to do." Helen got to see her young daughter officially open Ranscombe Farm Reserve: the botanist, campaigner and academic, Professor David Bellamy, had been brought in to perform the ceremony, but instead asked the youngest person there to take his place and cut the ribbon, as a way of symbolically passing the care of the natural world into the hands of the younger generation.

Helen Morley is not only quick to acknowledge the massive support given by her "A Team" of volunteers, but also recognises that her success in laying out the blueprint for Ranscombe Farm Reserve is due in great part to the help received from botanist Ros Bennett. It is as if the people behind a nature reserve make up an ecosystem in themselves, each bringing particular characteristics and skills to create something that is greater than just the sum of its parts. Or maybe a better analogy is the game of *Jenga* with its stacked wooden blocks: making a tall tower relies on each layer of blocks being supported by the others beneath it. This may also be better than falling back on Isaac Newton's famous words: I think what is illustrated here is none of us needs to be a giant in order to help create something important. The creation of Ranscombe Farm Reserve happened because of the inputs of lots of different people with lots of different skills, no two people being the same but all being indispensable to the final outcome.

When I came to work at Ranscombe Farm in 2011, I came to a nature reserve which was the sum of all the work of the people and organisations I mention above, and many more. Helen Morley had departed to go to art school, and her successor was Ben Sweeney, who took what Helen had

Some members of the 'A Team' of practical conservation volunteers, originally established by the first warden, Helen Morley, and still essential to the success of the reserve today. Photograph: Ben Sweeney/Plantlife.

achieved and ran with it, bringing to bear his own in-depth knowledge of the area, a sharp intellect and a keen imagination for what the site could and should be. In the six years since the reserve had been opened, it had gone from one small field known for its exceptional wild plants and where there was periodic, targeted work to conserve these plants, to a major new nature reserve. Work with the tenant farmer meant that the Cuxton Weedy Field was no longer an isolated island of rare plants: fields and field margins the length and breadth of the reserve were being managed for the kind of rare cornfield flowers for which Kitchen Field had once alone been famous, and to spectacular effect. Broad-leaved cudweed, once limited to Kitchen Field (and to only another half-a-dozen sites in the whole of Britain) was popping up all across the chalky field margins, joined by Venus's-looking-glasss, narrow-fruited cornsalad *Valerianella dentata* and other wild plants of arable farmland. Meadow clary had been given room and encouragement to multiply and spread, and rough mallow was popping up in other spots across the reserve.

This new reserve included 100 hectares (nearly 250 acres) of woodland which sheltered populations of plants known to be under threat in Britain as a whole: these included shy and diminutive fly orchids *Ophrys insectifera* as well as big and bold-as-brass lady orchids *Orchis purpurea*.

Just as important, perhaps, were the commoner woodland plants that bring joy to a spring day: yellow archangel as originally spotted by Thomas Johnson, sheets of bluebells *Hyacinthoides non-scripta*, scattered early-purple orchids *Orchis mascula*. Within a few years, this woodland – most of it a designated Site of Special Scientific Interest (SSSI) – would be brought into favourable conservation status thanks to Helen's and Ben's hard work. Ben recollects the state of the woodland before Plantlife started an active programme of coppicing: patches of meadow clary had been lost under scrub, there was little variety in the woodland flora, and populations of man orchid and lady orchid were just about hanging on. Ben describes the early stages of woodland management work as being "fairly large scale, and maybe on the face of it, quite brutal", focused as it was on hard-to-reach parts of the reserve and using heavy machinery. But a heavy hand was necessary because the previous lack of management would, in Ben's words, "have had considerably greater impact on biodiversity than the joyriding and car dumping that occurred at the same time" and because getting many of the hardest-to-reach coppice stands out of the way early on allowed various access improvements to be made to the site in the safe knowledge that these wouldn't too much affected by future work in the woodland.

The reserve also included areas of flower-rich grassland: the tiny patch where my sister and I had played as children – and where I had compiled my plant list as a sixth-form botanist – was being carefully managed, and retained its clustered bellflowers *Campanula glomerata*, burnet roses *Rosa spinosissima* and harebells *Campanula rotundifolia*. This lMill Hill grassland now had a much bigger sibling: the 23-hectare (60-acre) Brockles Field, bought with funding from the Heritage Lottery Fund and which, it was hoped, would eventually become as full of downland flowers as its smaller relative.

Ranscombe Farm Reserve was also decidedly still a real farm. Andrew Lingham, whose father, Peter, had taken the tenancy of the arable farmland in the 1970s, was still growing arable crops – mainly wheat, barley, and oil-seed rape – and fattening a few cattle in one small field. He remained an essential presence during my time at Ranscombe Farm Reserve, making it possible not just to cultivate ground at the scale needed to really benefit the site's rare cornfield plants, but also to bring in the grazing animals necessary to manage the increased area of wildflower grassland. This was not play-farming, though, but commercial management: yes, it used the environmental funding schemes available to all farmers, but it also produced crops of grain and other produce, and the cattle went to slaughter for meat. Whether Andrew thought having Plantlife as his landlord was a good thing, I would not like to say. But for me, Andrew was and is one of

Broad-leaved cudweed: Ranscombe is the most important British site for this very rare plant. Photograph: R I Moyse/Plantlife.

Burnet rose - a scarce plant in Kent, but still at Mill Hill.
Photograph: R I Moyse/Plantlife.

the good guys: decent, straightforward, extraordinarily hard-working, and caring about producing good food in a healthy environment.

The land that finally became Ranscombe Farm Reserve had grown from a small patch where rare plants had hung on for hundreds of years against a rising tide of losses and extinctions in the wider countryside. Now it was a wide, mixed landscape where wild plants were being positively encouraged to swim hard and fast against that tide. It was also no longer a hidden secret known only to a small band of botanists, but a place where plants were there to be seen: free to visit and criss-crossed by some 10 miles (16 km) of footpaths to take you from field to woodland to meadow and put you face to face with these amazing survivors. Anyone willing and able to visit could follow in the footsteps of three centuries of botanical explorers, stand where Darwin's tutor once stood, and see the direct descendants of the plants he saw.

It is this landscape with its plants and animals, fields and woods – but most especially the plants – which I want to turn to now. In the last couple of chapters, I have given an outline of the human story of Ranscombe Farm Reserve, but it is just an outline and there is a lot more that could be said about, for example, the work that it took to create and develop the reserve. After all, just in the time I was there, we spent some £750,000 of grant funding secured from the Heritage Lottery Fund, plus almost as much again from other sources, and many hundreds of days of amazingly good work were put in by scores of volunteers (including some, like the regular mid-week team of practical volunteers, who turned up again and again in all sorts of weather). But from this point, it is for the wild plants and animals to tell their stories: it is time to put them in the spotlight and for the humans to take a supporting role.

I am going to start with the cornfield weeds: the mainly annual wildflowers associated with cultivated farmland that earned Kitchen Field the name of the Cuxton Weedy Field. These may not be the plants which originally brought Ranscombe Farm to the notice of botanists, but they are undoubtedly the group of plants for which the reserve is now best known and for which it is most important as a place for plant conservation.

Chapter 3 - In the cornfields

I used to give a lot of talks to local community and social groups, trooping out with laptop, digital projector and screen to local gardening societies, Women's Institute branches and natural history clubs, and talking about Ranscombe Farm and its plants. Afterwards, I would often be asked about visiting the reserve: if people asked me the best time to visit Ranscombe Farm Reserve, I would normally suggest going sometime in the period between the end of May and maybe early August, with probably June to mid-July being the top time to see the really good stuff. Because this is when the cornfield weeds tend to be at their peak. If you are seriously into your plants, you can come and get your head down and your hand-lens out and look for some of the reserve's super-rarities, or you can just enjoy the parade of colours and textures and sizes and shapes of the different flowers.

I'm calling these plants 'cornfield weeds', but I did spend some time wondering whether it was the right term to use. It sounds a bit negative, 'weeds' being plants which we are expected to disapprove of, though if I call them 'wild plants of arable farmland', it just sounds a bit dry and technical. I like to think 'cornfield weeds' at least sounds a bit more friendly and old-fashioned, like something out of an old Ladybird book. Whatever term we go for (and I am very happy to stick to 'cornfield weeds'), these are the plants for which Ranscombe Farm Reserve is now nationally – and probably internationally – important: the mainly annual wildflowers associated with cultivated farmland. In other words, the wild plants of those fields used to grow wheat, barley, rye or other cereals, or other annual crops such as flax or oil-seed rape, where regular ploughing or similar cultivation is used to create the bare soil into which the crops can be sown: hence 'arable', which originally described land able to be ploughed or tilled.

Although the term 'weed' carries all sorts of baggage with it, these plants really are weeds in the sense of being unwanted interlopers in crop fields, able to take advantage of the whole process of arable farming. After all, if you are a farmer and trying to make a good job of it, then you really want your crop to occupy as much of your land as possible, and ideally you want the crop plants to be getting all the available nutrients, water and sunlight. You don't want other plants sneaking in and using some of the resources that might otherwise go towards growing food. But when you cultivate an arable field ready to be sown with wheat, or barley, or oil-seed rape or some other crop, you make a nice, bare seed bed where all sorts of plants can grow. Weed seed may blow in from elsewhere, or it might be lying waiting in the soil. It might even be hidden amongst the seed you've saved from last year's crop, so that you accidentally plant it at the same time that you sow the crop seeds.

A heavy load of weeds can really affect how much usable produce you get off the land, and farmers have struggled for centuries against the weeds growing amongst arable crops. The tide only really started to turn in favour of farmers in the late 19th Century, when improved seed-cleaning techniques meant that there was less chance of weed seeds lurking amongst the grain they were about to sow. Then in the 20th Century, chemical herbicides were developed, as well as varieties of cereal that could be sown much more densely and would grow closely together to deny weeds any space. As a result, cornfield weeds which were once commonplace in the British countryside went into steep decline and are now amongst our most rapidly disappearing plants. Some do still remain widespread and familiar, like the common poppy *Papaver rhoeas*; others, like the cornflower *Centaurea cyanus*, are still in our minds though long vanished from the countryside. Many more of the less showy species have gone from both our landscape and our memories. But, for some reason, Ranscombe Farm managed to hold onto many of the cornfield weeds that were rapidly disappearing elsewhere, and now they are thriving on a site which remains, as it ever was, a working arable farm.

The diversity of cornfield weeds at Ranscombe is astonishing, particularly when you understand that none of them (bar one or two very special exceptions) were deliberately put there. Rather, they are there because of their own ability to survive and grow and spread. There are, for instance, five different species of poppy, each of which differs from the others in its size, colour and behaviour. The common poppy (sometimes called field poppy or corn poppy) is the most abundant of these, and is probably one of our most recognizable wildflowers, with its four shining, deep-red petals, each dabbed with an inky black patch at the base. Common poppies have a fairly long flowering period from May to August, and can on occasion be super-abundant: when you see acres of farmland turn brilliant red, it will be common poppies that are responsible. Every so often, this summertime spectacle happens at Ranscombe Farm, to the delight of local photographers; more often, though, plants are to be found scattered amongst other cornfield flowers as a sort of patchwork. Almost as frequently encountered at Ranscombe Farm are opium poppies *Papaver somniferum*. With their waxy, grey-green leaves, flowers that vary from purple to almost completely white, and fat and voluptuous seed-pods, opium poppies are often grown as garden plants.

Opium poppies are, of course, the source of both legal and illegal drugs, not just opium but heroin and the important painkillers, morphine and codeine. Cultivation of the plant for these drugs is largely confined to

The five poppy species found at Ranscombe Farm Reserve:

Opposite, above: common poppy.

Opposite, below: opium poppy.

This page, above left: prickly poppy.

This page, above right: rough poppy.

This page, below left: long-headed poppy.

Photographs: R I Moyse/Plantlife.

warmer climates than Britain's, but the poppies are increasingly now grown and harvested under licence in Britain for medical use. This does not, though affect the opium poppy's status as a welcome and true British wildflower (and neither does it mean there is any problem with growing it in the garden). Opium poppies are also where we get the tiny, tasty poppy seeds which are used so much in Asian and European cookery.

A less frequently encountered poppy at Ranscombe Farm Reserve is long-headed poppy *Papaver dubium*, which only occasionally pops up in arable fields or other disturbed spots near the reserve's main entrance. The 'long head' of the plant's name is its seed-pod, which is long and slender and not the distinctive squat, urn-shape of the common poppy: this difference is obvious even in a newly opened flower. But even from a distance, the flowers look distinct and different from common poppies, and make you realise how many subtle variations can be played on a basic theme of scarlet petals: in long-headed poppies, the flowers have a washed-out look compared to the rich red of common poppies. They also lack the black base to the petals which most common poppies have and which give such a strong emphasis to the centre of their flowers.

The shape of the seed-pod is also the most distinctive feature of both prickly poppies *Roemeria argemone* and rough poppies *Roemeria hispida* Prickly poppies are slender and delicate and turn up in small numbers in fields on thin, chalky soils where they flower for a short period in early summer. Their orangey-red petals are so narrow that they stand wide apart in a fully opened flower, and they are followed by seedpods which are long and thin and dressed with stiff bristles. Rough poppies can be found in similar places, though often slightly later in the season, and their flowers are different again: small, neat and a deep crimson colour which is startingly different to the scarlet of common poppies. The seed-pods that develop from the rough poppy flowers are rounded and egg-shaped and densely bristly. Often, it is just as well the seed-pods of these two species are so distinctive, because it can be all you have to identify the plant. Both prickly and rough poppy flower in the morning and then only for a few hours: Prickly poppy flowers are particularly short-lived, with the petals dropping about lunchtime or even earlier if the weather is hot. So if you happen to be doing your plant-hunting in the afternoon, the most you will often find is a plant with some ripening seed-heads and some buds which may open tomorrow or the day after – and if there are a couple of shrivelling petals lying on the ground beneath, then that's a bonus.

There is a reason why prickly poppy and rough poppy make such a short show of flowering. It is not just to frustrate the botanist who can't get up in the morning, but part of a smart fail-safe process to ensure pollination and the production of seeds. Though both prickly poppy and rough poppy (and long-headed poppy, too) have showy flowers to attract insect pollinators, they are actually entirely capable of pollinating themselves. The anthers inside each flower, which produce the pollen, will start to open while the flower is still a tightly-closed bud, and long before any pollinating insect can get a look-in. This means that they are not dependent on pollination by insects in order to produce new seeds – of course, the flowers do open up and they might well get pollinated by the local bees or other insects, but if the insects choose to ignore them, then it's not a total wash-out and some seed can still be produced. As a result, prickly poppy and rough poppy flowers do not have to hang about waiting for an insect to visit and can crack on with making the next generation. In contrast, common poppies have to be cross-pollinated by insects and so their flowers tend to stay open till later in the day to maximize their chances. It's a riskier process for common poppies not being so self-reliant, but they do have the potential benefit of a bigger reward, as cross-pollinated flowers tend to produce more seeds. So there is a trade-off between going for smaller but more regular wins or taking a bigger risk for a potentially much bigger reward. You can't directly ask a poppy about its attitude to risk, but their behaviour gives it away: give a common poppy a few quid, and they'd buy a lottery ticket; a prickly poppy would tuck it into Premium Bonds.

Poppies are amongst the most obvious of cornfield weeds, but there are a lot more plants which like the regular disturbance that comes from arable farming. Some are common and widespread species that can turn up on almost any piece of disturbed ground, like the sprawling scarlet pimpernel *Lysimachia arvensis*, with its red, star-shaped flowers, or the pale green and decidedly modest annual mercury *Mercurialis annua* which can be abundant on the dry, chalky soils. Others, for reasons that are not clear but perhaps have to do with seed-dispersal ability, are fussier and tied more tightly to real farmland, and less able to make a home in town or garden. One of these is Venus's-looking-glass, a gem of a plant which can be hard to spot, being only around a hand-span high. But when found, they are unmistakable, bearing the most beautiful, deep-blue, five-petalled flowers above oval, ripple-edged leaves. Even in fruit, they remain attractive, each seed-pod being long and three-sided, with the sepals from the flower forming a kind of ruff at the tip: they always remind me of old-fashioned sweeties wrapped in a twist of paper. They should also get some sort of prize for best-name-for-a-cornfield-weed: the 'looking-glass' apparently

Stinking chamomile grows is super-abundance in some fields, and can fill the air with its pungent scent on hot summer days.
Photograph: R I Moyse/ Plantlife.

comes from the polished and reflective appearance of the individual seeds, but making it specifically the mirror of Venus seems sheer poetry.

Most plant names are more down-to-earth. Stinking chamomile *Anthemis cotula* for one: the plant smells unpleasant and looks like a chamomile – it has feathery foliage, and the flowers are built on the typical fried-egg pattern of a daisy, being yellow in the middle and white round the outside. The strong smell of stinking chamomile comes from aromatic oils produced by the plant, and they really are powerful: I know from experience that too much handling of the plant can make your skin tingle and smart. Stand in a field where there are thousands of plants in flower, on a really hot day, and the evaporating oils not only scent the air but can be so strong that your eyes start to water. And in some years, particularly in Longhoes Field and Kitchen Field at Ranscombe, there may be tens of thousands of plants, massively outnumbering those other chamomile-like species which tend to be so much more common in the wider countryside.

You will find both scentless mayweed *Tripleurospermum inodorum* and scented mayweed *Matricaria chamomilla* at Ranscombe farm and both are almost identical in appearance to each other and to stinking chamomile. The proper botanical differences between the three species depend on close examination: to pick out stinking chamomile, you break open one of the 'fried-egg' flowerheads, which consist of hundreds of tiny, individual flowers, and look for slender, green scales between these flowers. This kind of close examination requires a hand-lens for magnification, and a bit of experience. Fortunately, though, the plants do live up to their respective names, and a good rub and a sniff of the flowerheads often does the job: scentless mayweed doesn't smell at all, and scented mayweed has a pleasant, fruity scent quite distinct from the acrid odour of stinking chamomile.

A surprising number of plants are so similar to each other that proper identification relies on a close-up look with a hand-lens – or, at least, according to the text-books. Take the cornsalads, for example, which you may be familiar with, as cornsalad is grown and eaten as a salad vegetable, and sometimes turns up in those bags of mixed salad leaves in the supermarkets. The species which is grown as a vegetable is common cornsalad *Valerianella locusta*, but there are several other species which grow wild in Britain, and you can tell that no-one really worried about telling them apart until botanists started looking at them, because their English names are strictly functional and turn on fiddly little details about the fruits. So there is narrow-fruited cornsalad *Valerianella dentata*, keel-fruited cornsalad *Valerianella carinata*, broad-fruited cornsalad *Valerianella rimosa* and hairy-fruited cornsalad *Valerianella eriocarpa*. No real poetry there. Fortunately, the potentially tricky problem of identification is avoided at Ranscombe, as (so far at least) only two cornsalad species have been found: common cornsalad, which forms rather dumpy, untidy mounds, with heads of tiny, pale-blue flowers in early summer; and narrow-fruited, which is slender, upright and tidy-looking, with slim leaves and small heads of nearly white flowers. Easy to overlook, narrow-fruited cornsalad has a bearing which is almost military in its neatness and the symmetry of its branches.

Close-up examination of seeds can also be important when trying to pick apart the different species of fumitory, though there's a little less sheer botanical functionality in the names of the species found at Ranscombe: common fumitory *Fumaria officinalis*, dense-flowered fumitory *Fumaria densiflora* and fine-leaved fumitory *Fumaria parviflora*. The English name is said to derive from the Latin *fumis terrae* – the smoke of the earth – from some ancient association perhaps arising from the appearance of the grey-green foliage or the smoky smell of newly unearthed roots. They are

Three fumitories. Top left: common fumitory. Top right: dense-flowered fumitory. Bottom: fine-leaved fumitory. Photographs: R I Moyse/Plantlife.

unusual-looking plants with divided leaves and elongated, almost tubular flowers, and it is the shape and size of the various flower parts, and of the single-seeded fruit produced by each flower which are crucial in telling the different species apart (though the much smaller, white flowers of fine-leaved fumitory are a give-away – the other species have larger, pink flowers).

The poppies, fumitories, cornsalads, and the others I have described above, are just a few of the cornfield weeds to be found at Ranscombe. They are also all plants that are gradually vanishing from the wider British landscape, some a lot faster than others. Some of them are facing a real risk of becoming extinct in Britain, and so find themselves on British or English 'Red Lists'. Red-listing is a formal way of identifying the risk of extinction faced by an individual species at a chosen scale, whether this is an individual country, a whole region (such as Europe) or even world-wide. It may sound dry and staid – dispassionate when passion is surely what is needed to tackle biodiversity loss – but it is a very useful tool to understand what is happening to species and why, and to decide where effort might most be needed.

Red-listing for plants takes into account a whole range of factors, in a process that was developed by the International Union for Conservation of Nature (IUCN). To start with, it is important to know how rare a species is: perhaps it has been reduced to a small number of individuals, or it is found only in a small area. But it is more important to know what is happening to that species: is the population stable, or even increasing, or is it in decline; and, if it is in decline, how rapidly is it disappearing? Finally, it is essential to know in how many places the species can still be found: if all the eggs are in one basket, then the risk of extinction is much higher. You can then decide, given enough information, not just whether a species is at risk of extinction at the scale you are dealing with, but also how severe that risk is. The IUCN method puts threatened species into one of three categories: in order of increasing threat, these are Vulnerable, Endangered or Critically Endangered. Beyond Critically Endangered, there is only Extinct or, for species lucky enough to be maintained in captivity or cultivation, Extinct in the Wild. Those species which seem to be doing OK are categorised as Least Concern. And a species which appears to be in serious decline, but which is not doing quite so badly as to be considered Vulnerable, may be classified as Near Threatened. It is worth bearing in mind, though, that a plant can be rare without being under threat of extinction, so that, for example, Breckland thyme *Thymus serpyllum* is rare in Britain and not found outside the Breckland landscape around Thetford in Norfolk, but its populations are fairly stable and in the British Red List it is given as being of Least Concern. By contrast, prickly poppy, one of Ranscombe's poppy species, remains relatively widespread but has undergone such a substantial, recent decline in Britain that it is classified as Vulnerable.

By looking at the Red List status of the plants at Ranscombe Farm Reserve, we can see just how important the site is for cornfield weeds. Of twelve species found at Ranscombe Farm which are on the Red List for Great Britain, eight are plants of arable farmland; four out of five species in the Near Threatened category are also cornfield weeds. Stinking chamomile and fine-leaved fumitory are both classed as Vulnerable in the British Red List, while narrow-fruited cornsalad is Endangered. Also Endangered is broad-leaved cudweed, one of those species that visiting botanists started to take note of in the early 20th Century, when it was already starting to disappear from the countryside. In some ways, you do have to be a bit of a plant-lover to even spot broad-leaved cudweed; certainly, no-one could honestly say that it is exciting to look at. It is low-growing, grey-green and woolly, and with tightly clumped, spiky heads of tiny flowers that never really open, making it easy to overlook and, where it pops up in a footpath, as it sometimes does, easily trodden on (though, in keeping with its self-effacing appearance, it doesn't seem to mind this terribly much). But it is one of Ranscombe's most important plants: there are fewer than ten remaining sites for broad-leaved cudweed in the whole of Britain, and Ranscombe Farm Reserve is the most important of these, with many thousands of plants (the vast bulk of the entire British population) present every year.

In addition to these, ground-pine (Endangered) is another cornfield weed found at Ranscombe which is at risk of extinction in Britain, while the tiny dwarf spurge *Euphorbia exigua* and the tall and elegant grass, rye brome *Bromus secalinus*, are considered Near Threatened. It is an amazing list for a single site, and it is complemented by other plants which are less threatened, such as the spikily-scented corn mint *Mentha arvensis*; blue pimpernel with its flowers of an intense, deep sky-blue; the pink and perky small toadflax *Chaenorhinum minus*; the prickly-leaved and pale-blue-flowered bugloss *Lycopus arvensis* (sometimes called small bugloss to distinguish it from its bigger, brasher cousin, viper's-bugloss *Echium vulgare*); and the two fluellens – sharp-leaved fluellen *Kickxia elatine* and round-leaved fluellen *Kickxia spuria* – each with tiny yellow-and-purple snapdragon flowers on their slender, trailing stems. In fact, it's no exaggeration to say that Ranscombe is probably the single best place in Britain to see cornfield weeds – not only because of the overall number of

Narrow-fruited cornsalad. Ranscombe Farm Reserve has one of the largest British populations of this endangered cornfield weed.
Photograph: R I Moyse/Plantlife.

species, the number of rare species, and the large size of the populations of some of the rare species, but also because most if not all are easily found near footpaths, or even, in the case of broad-leaved cudweed, occasionally in the path itself.

It's worth reiterating that (with a few exceptions, such as the perennial corn mint) these cornfield specialist weeds are all annuals. That is, their entire life-cycle is completed within a year: seed germinates either in autumn or the following spring, depending on the species (though some hedge their bets and can do either), and the new plant then grows, flowers, fruits and dies before the next winter. Some – like poppies – die because they have made all the flowers they can; others – scarlet pimpernel is a good example – will keep growing and branching and making more flowers till they are killed by the frosty weather of the coming winter. Either way, that's it: any one plant gets at most a year of growing, after which it's gone and all that remains are the seeds it has managed to produce during its short life. The population's survival depends on those seeds being able to make a go of it. It's a chancy lifestyle for any wild plant because those seeds will need to find a patch of bare ground, open to the light, where they can germinate and grow without being shaded out by other plants. In ploughed farmland, these conditions are available in super-abundance, but farming is a human invention, and big areas of freshly turned-over bare ground just don't turn up very often in nature.

Despite bare ground being a rare commodity in nature, this lifestyle of being able to rapidly exploit patches of bare ground is one of the three key plant strategies identified in the important work of Professor J. P. Grime of Sheffield University. Technically, it is called being a ruderal plant, with the word 'ruderal' apparently coming from a Latin word for rubble. The two other key strategies which plants use to survive and succeed are to be competitive by outgrowing all the other plants present and acquiring most or all of the available resources (water, light, nutrients); or, alternatively, to be a stress-tolerator able to grow, even if only slowly, in very difficult conditions. So, for example and in general terms, a beech tree would be an example of a competitive plant: by spreading its branches high above the ground, it can shade out other vegetation and claim all the light, while roots below ground are able to claim the water and nutrients. Conversely, a plant like, say, wild thyme *Thymus drucei* growing on chalk downland has many of the characteristics of a stress tolerator: it is able to grow on very shallow, well-drained soil, in hot sun, where other plants would shrivel up and die.

Ruderal plants – including our cornfield weeds – take the different approach of avoiding too much competition by getting in quick where the ground has been disturbed and the existing vegetation destroyed: using this rapid-response and having a short lifespan, they are able to flower and seed before other, more competitive plants even get started.

I like to think of these annuals as the plant equivalents of Del-boy Trotter in the BBC sit-com, *Only Fools and Horses*. Bear with me in this. In what you might call this retail-sales-based model, you can see competitive plants as being like the big supermarkets: they need a good market to exploit, such as bread, toilet rolls or other staples, and can dominate the market for these mass-purchase goods, pushing out small competitors. Stress-tolerators might, in contrast, be represented by a specialist second-hand bookshop, selling to a small clientele that the large traders are not interested in, and able to scrape by on low turnover and small profits. Whereas the ruderal plants are like Del-boy, who's used to spotting an opportunity to open his suitcase and sell a few dodgy watches before the police appear and chase him away: it's all about a fast turn-around and grabbing as much as you can before conditions change.

Outside of the farmed landscape, significant events that disturb the ground and create gaps in existing vegetation are not very common. In the wild, such events tend to be localised and unpredictable. For example, trees might fall, and in doing so create a gap in the woodland canopy and a bare patch of ground where the root-plate has been up-ended. Flood events on rivers, or coastal erosion, also create bare ground. Before Britain lost most of its larger wild mammals, wild boars may have grubbed and rooted around, or congregations of wild cattle or horses churned up the ground with their hooves. But even where animals of this sort are still present, this sort of occurrence will tend to be patchy. Not only that, but any disturbed ground will fairly quickly vegetate over and won't stay bare for very long. With suitable conditions for seed germination being scattered in space, and pretty short-lived where they do occur, ruderal plants need to have clever methods for finding those bits of bare ground where and when they happen.

Some do this by having seeds that move through space. Dandelions *Taraxacum* spp. are a classic example, with each seed having a parachute of hairs which allows it to be carried away on the wind. The principle behind this is playing the odds: if you make enough seed and send it out across the landscape, some is likely to land in a suitable spot. But, interestingly, most of the rare cornfield weeds don't do this; instead, they tend to have seeds which just drop within a few feet of the parent plant. Some might be able to throw their seeds a short distance: the distinctive seed-pods of poppies are designed as a kind of pepper-pot, with holes around top so that the seeds are thrown out when the wind shakes the stems;

Blue pimpernel growing alongside scarlet pimpernel in Kitchen Field.
Photograph: R I Moyse/Plantlife.

field pansies *Viola arvensis* go further and have seed-pods which open explosively, throwing out the seed. But neither are ever going to send their seeds very far. So instead, most cornfield weeds don't send their seeds through space to find suitable habitat – instead they use the fourth dimension and send seeds through time. I'm not claiming that somehow plants have invented time travel: what these plants have is not really a time-machine but a mechanism to keep the embryo plant alive within the seed for a very long period. All the same, it does mean that cornfield weeds effectively treat the universe as four-dimensional space-time in much the same way as Einstein did. If the conditions a plant needs for germination and growth are scattered patchily through both space and time, then sending seeds some*when* else makes as much sense as sending them some*where* else. This is one of the weird and exciting things that plants do which still makes my mind boggle: it took humans till the 20th Century to understand the universe in this way, but plants had for millions of years been acting as if they knew it all along.

The key to achieving this effective travel through time is that old staple of sci-fi space travel: suspended animation. The embryo in the seed must be able to survive, sealed safely into a tough case, for many years. In the classic New Naturalist volume, *Weeds and Aliens*, Sir Edward Salisbury states that poppy seeds can live at least 80 years, possibly more than a century, in the soil. From my own experience, it does seem that a twenty-year life in the soil is a piece of cake for many cornfield species. But there's another old staple of sci-fi, which is that the astronauts survive their years in suspended animation, only to wake to some massive disaster which finishes them all off (except for whichever star has their name above the title on the poster). In a similar way, plants adopting the time-travel strategy would put the next generation in danger if all the seeds germinated together. You might have hundreds of seeds which have sat in the soil for decades waiting for a sudden episode of disturbance to create the bare ground they need. What if they all respond at the same time, and germinate, only to find some other unexpected event wipes them all out? As a fail-safe, there needs to be a mechanism to stop all the seeds germinating together, otherwise all their eggs are in one basket. This was brought home to me by the amazing behaviour of blue pimpernel.

Blue pimpernel is a species mainly found in Southern Europe, but which creeps into the southern parts of Britain and is regarded as Nationally Scarce here. In general form, it resembles the much more common scarlet pimpernel, and confusion is added by scarlet pimpernel having a variety with blue flowers, so that – again – careful examination with a hand lens is needed to be certain that what you are looking at really is blue pimpernel. At Ranscombe Farm, it occurs regularly only in one field – Kitchen Field –

at the western end of the reserve, and then only on chalky soil in the upper part of a south-facing slope. You can find it most years, though numbers vary and generally there's just a handful of plants to be seen: half-a-dozen or so is a regular count. But one of the things about dealing with plants that grow anew each year from seed is that you never know what's going to happen next, and there was one year when blue pimpernel just went berserk, producing an estimated 7,000 to 8,000 plants. Like scarlet pimpernel, blue pimpernel doesn't just produce a set number of flowers and then die; rather, it keeps growing, branching and making new flowers till it succumbs to the cold weather in autumn. So those 7,000 or 8,000 plants must have produced literally millions of seeds and dropped them into the soil. Yet that massive display of plants has never yet been repeated: in the six or seven years since, the maximum number of blue pimpernel plants in any year cannot have been more than a couple of dozen.

This has obviously been a bit disappointing – apart from anything else, blue pimpernels are cracking to see – but it makes sense for the plant. If a plant population makes millions of seeds, the worst thing it can do is have all these seeds respond together, especially in the case of a plant that is killed by the cold: one unexpected frost in late spring, when the seedlings have already popped their heads up, and you've blown it. It's clear that, like

Reliably separating scarlet pimpernel and blue pimpernel can be tricky, as scarlet pimpernel can also have blue flowers. Check with a hand-lens - scarlet pimpernel petals (left) have a dense fringe of hairs, and the petals almost hide the green sepals beneath. In blue pimpernel, the fringe of hairs is sparse and the petals spaced more widely apart.
Photographs: R I Moyse/Plantlife.

many species that rely on a well-stocked bank of seed in the soil, blue pimpernel has a time delay built into its seeds. Whether this is a physiological system – some clever biochemistry that delays development – or something mechanical like a hard seed-coat which needs to abrade before germination can occur, or some combination of these or other mechanisms, I cannot say. But it makes sense for the plant to have a system to delay the germination of some – perhaps most – of the seeds produced in any given year.

And perhaps this delay, together with the long-term persistence of seeds in the soil, is a common feature of the weeds of arable farmland because otherwise these plants wouldn't survive the efforts of generations of farmers to eradicate them. The wild plants of arable farmland are today amongst the fastest declining plants in Britain, and this owes a lot to modern farming techniques and especially the use of herbicides. But without access to chemical herbicides, earlier farmers would still have had methods to clear weeds from fields. No farmer would have wanted weeds in their crop, as they take up space that could be growing food instead – today it can make for a substantial financial loss, and in the past it could have meant starvation – and before herbicides were available a weedy field might be put down to grass for a year or two before ploughing it again. Or they might make a stale seed bed by ploughing once, and then letting the weeds germinate before ploughing again to kill the newly-sprung weed seedlings and then finally sowing the crop. But neither of these is going to be proof against a plant clever enough to have seed which not only survives for years in the soil, but which also isn't foolish enough to jump at the first chance of germinating.

This is also, in part, how Ranscombe Farm came to be such a superb site for cornfield weeds: the plants did it by being smart enough to survive. Though they were helped by two things. One of those things was Plantlife taking on the management of Ranscombe Farm Reserve in 2005. Plantlife wanted to ensure that Ranscombe Farm became a truly exceptional site for wild plants, and Plantlife has an ethos that focuses on letting wildflowers be properly wild.

This may be worth dwelling on for a bit. It is very easy to take a bit of land with few plant species and sow it with a wide range of wildflower seed. Seed is easily purchased, lots of species are available, and the results are rapid. Once the flowers start blooming, lots of insects will move in. Bees and butterflies in abundance – result, right? Well, yes, assuming that you are only interested in the insects, and the plants are only there as a kind of substrate on which the insects can feed and grow. But it does nothing for plant conservation. You wouldn't, after all, think you'd achieved much for butterfly conservation by rearing a dozen species from caterpillars, then releasing the adult insects. If you want to secure a future for butterflies, you need to create and maintain habitats where they can feed and reproduce, and you want to ensure that patches of suitable habitat are connected in a way that allows the butterflies to spread, increase in number and colonise new parts of the landscape. You can casually release as many butterflies as you like, but all that is going to do is create competition with wild butterflies, possibly displace species already present, and confuse the picture for anyone trying to find out exactly what is happening with wild butterfly populations.

As with butterflies, so with wild plants. To really make conservation work for wild plants, you need to create the right sort of environment so that the plants are able to increase and spread through the landscape. And it's very hard to understand how to do this if you're not sure if a plant is where it is because the habitat is right and it has managed to naturally find its way there, or simply because someone has come out and planted it. This doesn't mean that there is no role for introducing plants or seeds to an area. Again, as with butterflies, it may be that there is no way that a particular species could ever naturally reach a certain area of land which might otherwise be suitable for it, or it may be that the urgent need for conservation means you cannot wait for it to happen. But deliberately bringing any new species to a site must always be the last resort and happen because other measures have failed or are impractical. Animal introductions are always carried out with care and respect because, well, they are animals and can suffer. Maybe it is because plants don't suffer in the same way that it feels OK to casually move them about. But it really seems to me that we should also be respecting the wildness of the species we are trying to conserve, and we should always give any species, whether plant or animal, the chance to survive and thrive as a truly wild-living creature.

So the Plantlife approach at Ranscombe Farm Reserve was to simply manage as much as possible of the arable land in a way which would allow the plants to do their thing. This meant taking into account three things: the plants are annuals which rely on the yearly cultivation cycle of arable farmland; chemical herbicides are one of the key factors in the decline of these species in the 20th Century; and some of the species also fail to respond well to the addition of chemical fertilisers. So effective management meant ploughing or otherwise cultivating the ground every year, and making sure no herbicides or fertilisers were applied.

But the other thing that helped the plants do their job of turning Ranscombe Farm into the best place to see cornfield wildflowers was that so many species had managed to hang on there in the first place. The land there has been farmed for centuries, and would have endured the rigours of the huge intensification of farming that followed the Second World War, and yet many species clung on while disappearing elsewhere. The book published in 1977 to mark the centenary of the now sadly defunct Rochester and District Natural History Society includes a walk through Ranscombe Farm where it mentions plants including cornsalad and blue pimpernel, and this at a time when it was a modern and entirely commercial farm. What made Ranscombe Farm so lucky? Well luck, or rather chance, may be important: after all, somewhere has to be the last location of any disappearing species. But I think the landscape itself may play a role, because old maps show that the basic pattern of tracks, fields and woodlands at Ranscombe Farm Reserve has remained unchanged for considerably more than 200 years. You can look at a map from the late 1700s, and, apart from a handful of hedges and tree-lines, it is the same place. Even a map from the mid-1600s shows virtually the same field pattern. One field, for example, called Cobham Field on the old tithe maps, has an odd shape: the northern edge, where it meets Cobham Woods, has a stepped outline. This strongly suggests that the original field was smaller and that a series of extensions has been added to it by clearing successive strips of the adjacent woodland. This must have happened over a number of years, as each newly-added strip was a different length and width to the previous one, so that the footpath that now runs along this edge follows a winding path around a series of more-or-less right-angled corners. As a result, the shape of the field is distinct and recognisable, and it is easy to see that it was already in place more than 300 years ago.

This consistency of field pattern is important, because field edges are the places where arable wildflowers were able to hide out during the intensive farming of the late 20th Century. With big farm machinery, it's hard to work tight against the edge and into corners, so it was the field margins that tended to be missed by the herbicide spray or fertiliser spreader. Where those field margins have stayed the same for hundreds of years, then they are likely to be a storehouse of buried seed, just waiting to go.

We've certainly seen things like this happen at Ranscombe Farm. When Plantlife took on the reserve in 2005, Kitchen Field, at the far western end of the site, was already well known for its rare cornfield weeds, which included broad-leaved cudweed, blue pimpernel and stinking chamomile, and for this reason it had been a Site of Special Scientific Interest since the mid-1980s. Elsewhere throughout the site, and especially on shallower, chalky soils, other rare plants could be found scattered on field margins.

Bringing those margins into the right sort of management led to an explosion in the cornfield weeds, as they found themselves released from the competition of a sown crop and no longer having to face the challenge of herbicides. Nowhere was this more true than in Longhoes, the long, narrow, sloping field close to the reserve's main entrance.

Longhoes, in common with most of the farmland at Ranscombe Farm when Plantlife took over, was a commercial arable field, sown annually with a crop and given all the necessary treatment to ensure that crop was as productive as possible. Then, shortly after the reserve was set up, Helen Morley, the site's first warden, spotted the give-away grey-green leaves of a couple of plants of broad-leaved cudweed in a corner of the field that had been missed by the herbicide-sprayer. Helen persuaded the farmer to leave the top margin of the field unsown and untreated the following year, and plants immediately returned, bursting up from long-buried seed, including big numbers of broad-leaved cudweed. So, as soon as it was possible to bring the whole field into similar conservation management – using financial support from the Higher Level Stewardship agri-environment scheme – this was done. Over subsequent years, the cornfield plants used their ability to spread a little distance to colonise the whole field, aided of course by the way in which soil is moved about as the ground is cultivated and turned-over. The numbers and distribution of plants in Longhoes vary – sometimes wildly – from year to year, as different species respond in different ways to the timing of cultivation and to the weather. But, in recent years, populations of rare plants in Longhoes including broad-leaved cudweed, narrow-fruited cornsalad and fine-leaved fumitory have numbered in the thousands; and, one year, there were an estimated two to three million broad-leaved cudweed plants in a field which had been just another piece of commercial arable land less than a decade before.

The resilience of cornfield weeds and their ability to bounce back from near oblivion, using their clever adaptations and time-travel abilities, is astonishing to witness. Watching the populations of different species wax and wane from year to year, sometimes turning up unexpectedly in a new place, sometimes vanishing for a year or two (though you know the seed is still waiting in the soil) is like watching a sort of soap opera where the plants are the characters. Fortunes change, new characters arrive, or an old character returns. Different plants may dominate in different years. Every so often there's a big cliff-hanger: will a late spring mean a population crash? Will the plants recover from a summer drought? And as with any good soap opera, you get drawn into the characters' lives. There are complications, frustrations and surprises, and there is a story going on, though not one with a pre-determined 'narrative arc' – we don't know where this is going or how it's going to turn out. But the plants, with their comings and goings, their failures and successes – most of all, their persistence – are telling something of the story of a place.

Which makes you wonder – what's their back-story? Every great character in a TV drama has a back-story which provides the key not just to their own behaviour but sometimes to the whole narrative. Cornfield weeds most definitely have a back-story, and it's one that needs to be illustrated by looking at a cornfield plant I haven't mentioned yet. Corncockle *Agrostemma githago* is a plant that used to be common and widespread in arable farmland. Tall – a good specimen can be three feet high – and with five-petalled, purple flowers each the size of an old, pre-decimal penny, it is a truly beautiful thing, but has undergone such a massive decline in numbers that it was, until recently, classified in Britain as Extinct in the Wild. Part of the reason it has suffered so badly is that its seeds cannot do the time-travelling thing. Unless very deeply buried, where the soil is dark and low in oxygen and decay is slowed, corncockle seeds must germinate quickly if they are not to rot and die. Sometimes seed does find its way deep into the soil, and sometimes this deeply-buried seed does get brought to the surface and germinate: there are occasional incidents of plants popping up during major construction projects, such as road-building, but they rarely establish a lasting population. It is deeply-buried seed which seems to have been the origin of Ranscombe Farm's corncockle population. A couple of decades back, Kitchen Field was deeply ploughed to try and bury – and so control – certain perennial plants which were starting to be a problem for the cornfield weeds, and it is this which is thought to have brought up some corncockle seed which subsequently germinated.

Given its head, corncockle can do quite well. Each flower produces around 22 or 23 big, dark seeds, and these germinate readily in autumn if there is bare, disturbed ground available. But leave cultivation to the following spring, and perhaps around 90% of the seeds in the surface soil will already have rotted off. Wait a year before cultivating the field, and pretty much all the rest will have gone. So, to keep going, corncockle really depends on the field being cultivated every autumn, for year after year. Now, beautiful though corncockle is, it was historically one of the most hated cornfield weeds. Because the seeds are large and heavy, they were very difficult to clean from cereal grain. The old method of cleaning seed after harvest was to winnow it: you put your harvested seed in a big, shallow basket and threw it into the air so that the breeze could carry off the chaff and any small weed seeds. The heavier cereal grain would fall back into the basket, and you would repeat the process until the grain had lost enough of the weeds and chaff and other bits of rubbish and could go to the mill. But the size and weight of corncockle seeds meant that they would just drop straight back into the basket with the cereal. Then, when the grain

Corncockle growing with common poppies in Kitchen Field.
Photograph: D R Moyse.

went to the mill for grinding, the corncockle seed would discolour the flour and make it unpalatable. The seeds are also toxic if eaten in enough volume, so that repeated consumption of contaminated bread could lead to a condition called githagism, which caused tiredness, weight-loss and stomach problems, and which could be fatal if the contaminated diet continued.

For this reason, farmers would do all they could to get rid of corncockle, and the very short persistence of its seeds in the soil should have made this easy: just put the field down to grass for a year or so, and all the seed would die and leave you with a nice, clean field. Except that the seed then sown would have been saved from a previous year's crop, and that might also contain corncockle seeds which had made it through the winnowing process. So, in sowing your new crop, you would be unable to avoid sowing corncockle as well. This was exactly how corncockle survived for centuries as part of the farmed landscape – the seed couldn't persist in the soil, but it had just the right characteristics to rope in farmers as unwitting and unwilling accomplices in its own survival and spread. Only when more efficient, mechanical seed-cleaning started to be employed in the 19th century did corncockle start to suffer, and it was already in decline at the beginning of the 20th Century; as farming techniques improved, corncockle was driven to the point of extinction.

We think of cornfield weeds as living in a perpetual struggle with the human management of the landscape, surviving through their clever abilities to respond rapidly to the opportunity offered by the cultivation of arable fields, and to send their offspring through time to find and exploit future opportunities. But the story of corncockle is not so much one of a plant fighting for a place in the farmed landscape, but more the story of a plant which achieved success by developing an intimate relationship with pre-industrial farming, and, indeed, with the farmers themselves. Corncockle was once widespread in the British landscape, and for centuries, but only because of arable farming and the hands of the individual farmers. But then it turns out that this is also true of the overwhelming majority of plants we think of as cornfield weeds, and it's a relationship that goes back to the very start of farming as a human endeavour.

Chapter 4 - Arrival of the weeds

Farming is often said to have started in the so-called Fertile Crescent, a stretch of the Middle East extending from the Nile Valley to the Persian Gulf, and taking in modern day Syria, Lebanon, Israel, Palestine, and Jordan, as well as parts of Egypt, Turkey and Iran. In practice, the domestication and cultivation of plants as food crops happened independently in a range of different places around the world; for example, potatoes and maize were domesticated by peoples in the Americas, coffee and millet in Africa, and soyabean and sugar cane in East Asia. But when we are talking of the kind of farming and the kind of crops we are most familiar with in Europe, then these do have their origins among the societies of the Fertile Crescent. It was some 10,000 years ago when people previously living as hunter-gatherers discovered that certain local wild grasses and other plants could be grown and managed and selectively bred to produce a larger and more reliable crop. The development of farming allowed larger populations to be supported by the same area of land and led to the development of settled (rather than nomadic) populations and the creation of the first cities. But the wild ancestors of the new domesticated crops shared their environment with a large number of other annual plant species, and these wild plants started to make their way, uninvited, into the new cultivated fields: finding these fields to be very much to their liking, they became the first cornfield weeds.

Over thousands of years, the farming way-of-life spread to the north and west, travelling around the northern edge of the Mediterranean and up into Europe, before eventually arriving in Britain around 6,000 years ago. For a long time there has been debate as to whether the spread of farming took the form of an actual migration of farming people, or if it was a spread of farming culture: that is, if it was simply the practice of farming that spread, as farming methods were learnt by existing populations from their neighbours. Genetic studies now suggest the former: farming people moved across Europe, mixing and interbreeding with the existing populations they met, and replacing the hunter-gatherer societies with new, settled societies.

The new farming communities, as they spread up into Europe, carried with them the seeds of the crop plants that had originally been domesticated and developed in the Fertile Crescent; amongst those crop seeds were the seeds of weed species, incidentally harvested with the crop and impossible to fully clean out with the winnowing techniques available at the time. This wasn't just corncockle, but all the different species of poppies, as well as narrow-fruited cornsalad, stinking chamomile, the two fluellen species and many, many more. Unlike corncockle, these other plants did not depend for their survival on being resown with the crop, but it must have helped them when they first colonised the new farmland environment – and it certainly provided them with the means of transport to parts of the world they wouldn't otherwise have reached. Today, the wild flora of the Middle East still includes all these species, though the area's long history of arable farming makes it hard to be certain exactly where any cornfield weeds had their original, wild homes. Nonetheless, there is some research on the wild origin of cornfield poppies which suggests, for instance, that prickly poppy was originally a wild denizen of Anatolia, the Asian part of modern Turkey at the western end of the Fertile Crescent. Long-headed poppy may have also first been wild in Anatolia, while common poppy's origins may well lay further east around modern-day Israel and Palestine.

Round-leaved fluellen - a weed of the first farmed fields of the Fertile Crescent. Photograph: R I Moyse/Plantlife.

At the same time that farming was being developed in the Middle East, lower sea-levels meant that Britain was still connected to mainland Europe across a broad front which stretched from modern-day Kent to Yorkshire. The ice sheet that had covered most of Britain north of a line from the Severn to The Wash had by this time long retreated so that Britain had been free of permanent ice for more than a thousand years. Plants and animals were spreading north and west from what is now mainland Europe as the climate warmed, and people were amongst them. These people were not farmers but Mesolithic (Middle Stone-age) hunter-gatherers, most of whom led a nomadic or semi-nomadic lifestyle, moving as their sources of food waxed and waned, or – in the case of the larger animals they hunted – as their food itself moved. The Mesolithic people were the same as us: the same species, with the same intelligence and same ability to communicate; they were just without the kinds of technology that developed later and therefore reliant on creating tools from stone, wood or bone. Other human species, different to us but still closely related, had been in Britain long before: the 500,000-year-old remains of *Homo heidelbergensis* were found at Boxgrove, 60 miles south-west of Ranscombe Farm, and the skull of Swanscombe Woman, an early neanderthal *Homo neanderthalensis*, shows that this human species was living just eight miles from Ranscombe Farm some 400,000 years ago. The first of our own species, *Homo sapiens*, arrived in Britain around 40,000 years ago, at about the same time that neanderthals were going extinct. But changes in the climate meant populations came and went over time, with successive advances by the ice sheets driving people out and back into mainland Europe. It is only with the arrival of the Mesolithic people some 12,000 years ago that we started our current history of continuous occupation of these islands.

These first Mesolithic settlers were smart and skilled enough to be able not just to find food and shelter in our previously unexplored corner of Europe, but to spread and increase in number, creating (crafting, more accurately, as the skill needed is tremendous) the tools they needed from the materials available to them, navigating without roads or compasses, and adapting their behaviour to suit local circumstances and the best food locally available. They would have been nomads by necessity, as hunting and gathering in our temperate part of the world, where the seasons are clearly defined, mean going to where you know food is to be found and knowing when to do so. They would also quickly have learned how to modify the environment to get prey animals to come to them, perhaps encouraging or maintaining open habitats to attract grazing animals.

Nomadic people don't leave big obvious signs like the remains of buildings, but we have good evidence of the Mesolithic occupants of Ranscombe. Work by long-time local resident, Ranscombe Farm Reserve volunteer, and hugely knowledgeable local amateur historian, David May, uncovered a substantial find of Mesolithic worked flint which had lain largely undisturbed for more than 6,000 years. This discovery was a happy accident: back in December 2011, part of Cobham Woods was being coppiced, and the vehicle bringing out the timber had churned up the surface of one of the existing woodland tracks. Dave has a good eye for detail, and he spotted some unusual-looking bits of flint in the disturbed and rutted mud: after he took home and washed the pieces, there turned out to be some 30 flint fragments that showed signs of being deliberately worked. Over the following week, Dave collected several hundred pieces of worked flint from the track, which has led to years of further work and exploration, often carried out with the support of the local community archaeologist, and together with other volunteers. Dave now has a collection of more than 15,000 pieces of worked flint – and that number is not a typo: it is a massive collection. Fourteen-thousand of these pieces are from the area of the original find, and include a good number of the stone tools that the Mesolithic people depended on – adzes, blades, pointed tools for hole-making, sharp edges for arrows and spears – though most of the pieces are the bits of waste flint left behind as a large flint nodule was carefully worked into one or more tools. The tools themselves always seem the most exciting pieces, at least to non-archaeologists like myself, and it is amazing to hold a flint adze that is more than 6,000 years old and to realise it has been shaped deliberately so that it sits comfortably in the hand.

But the waste pieces are also important and tell an amazing story. Some of the work by Dave, and the other volunteers who joined him, involved digging carefully planned test-pits which uncovered the original Mesolithic soil surface, where it was discovered that bits of waste flint were lying on this surface more-or-less where they'd first fallen. By carefully looking at the colour, texture and shape of the waste pieces, it was possible to see how some groups of these small fragments had come from the same bit of flint.

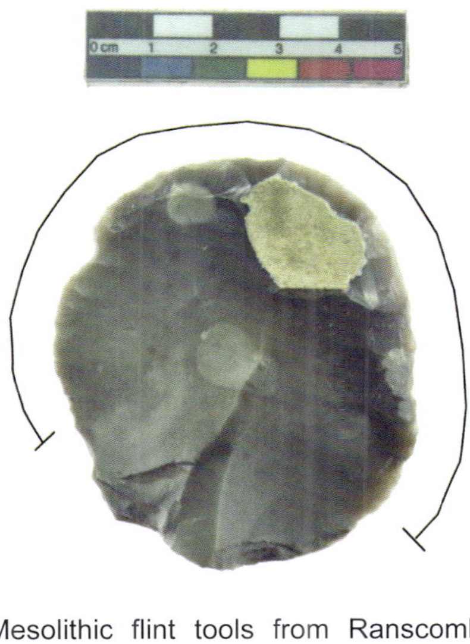

Mesolithic flint tools from Ranscombe Farm Reserve.
Left: an axe, shown from all four sides.
Above: an all-around or horseshoe scraper, with the long working edge indicated.
Photographs: David May.

So not only do we know where some individual workers sat, but some of the waste flakes could even be fitted back together to show the rough shape of the original flint nodule, and even the order in which bits were carefully chipped off to make the finished tools. For example, one group of four pieces, all found within an area of about four square-feet (well below half a square metre), fitted together to show that someone once sat on that very spot to work on making an axe before giving up and dropping the unfinished tool. Dave's refitting of flints is an astonishing achievement. Perhaps the nearest analogy would be if you were faced with a big pile of jigsaw pieces, each roughly similar in colour, belonging to an unknown number of different jigsaws, all of which have pieces missing. Plus you have no picture to work to, nor even any idea of the finished size or shape of each jigsaw. Yet, one refitted group which Dave was able to reassemble included 20 separate flint fragments.

The evidence from the Ranscombe Farm Mesolithic site, which has been studied by university academics – as well as being presented by Dave May at a national Conference of the Lithics Society – apparently suggests repeated occupation over a number of years, presumably as part of a nomadic perambulation through the wider landscape. But why Ranscombe, and what might it tell us about the site's plants and vegetation? The places where flints have been found are all in parts of the reserve which are wooded today, and the landscape would presumably have been largely wooded at the time: the Ranscombe Farm Mesolithic site is dated to around the time when woodland is supposed to have reached its maximum extent in Britain. But the location of the site on a slope on high ground above the Medway Valley does suggest a strategic position where you might be able to look down towards the river valley below to keep an eye on movements of wandering game animals, and perhaps also watch for other people moving up the river. Dave May points out that from the main Mesolithic flint site, it would have potentially been possible to see for two miles over an arc of nearly 270°, and on clear days the Thames would also be visible to the north – but none of this would have been possible if the view was entirely obscured by trees. Then again, the excavation work shows that the original Mesolithic soil was just sand and gravel, so perhaps this inhibited tree growth to the extent that wild grazing animals, or the Mesolithic people themselves, could maintain open, low-growing heathy vegetation. It has been suggested that Mesolithic people used fire as a tool to manage the landscape, and, while British broadleaved trees are very difficult to burn, it is far easier to burn areas of gorse, heather and the kind of heathy vegetation that develops on gravelly soils, and, in doing so, to slow or halt succession to woodland.

૨

By around 6,000 years ago, the expansion of the farming people and their culture finally reached Britain, marking the start of the Neolithic (New Stone-age) period here. The change from groups of hunter-gatherers to farming societies locks, from this distance, like a revolutionary cultural change, with the new, settled communities and big, stone monuments, but may at the time have looked more like a gradual evolution. After all, it is believed that Mesolithic people to some extent managed the wild herds of the animals they hunted, which may have included deer and wild cattle. There is even some evidence of trade between Britain's Mesolithic people and the Neolithic farmers of Europe, in the shape of cultivated grains found in coastal sediments dating from some 2,000 years before they would have first been grown here. This trade would have involved crossing the English Channel, as Britain had already been an island for around 2,000 years, but this was clearly not a major barrier and it appears that Neolithic people arrived in different parts of Britain at different times over a period of maybe hundreds of years, with their arrival in South-east Britain perhaps being the third wave of settlement.

A flint arrowhead from the late Neolithic or Early Bronze Age, found at Ranscombe by David May. Photograph: David May.

Ranscombe Farm Reserve includes a working arable farm today, and farming there certainly goes back many centuries, but did farming start at Ranscombe as far back as the Neolithic? It's impossible to say for certain, but we know that people were already living at Ranscombe when Neolithic people arrived and started to mingle with the existing societies. And it is clear that the Medway Gap – that notch in the North Downs cut by the River Medway and where Ranscombe sits – was an important, settled location from the early Neolithic onwards. Just a few miles south of Ranscombe is the cluster of long barrows and other stone monuments known as the Medway Megaliths, which includes such structures as Kit's Coty House, the Coldrum Stones and others. Evidence for how the Neolithic population affected the local landscape comes from a study conducted in the Medway Gap prior to the construction of the Channel Tunnel Rail Link in the 1990s This looked at the remains of snails found in soil deposits of identifiable ages, and used these to give an indication of the type of vegetation around the Neolithic monuments and other structures: different snail species have different habitat preferences, with some confined to woodland, others found in scrub and wood-edges, and still others associated with dry, open grassland. This showed that the early Neolithic landscape was heavily wooded – though with signs of disturbance and open ground – with an increase in open ground by the late Neolithic and through the Bronze Age. By the early Iron Age, around 2,800 years ago, the study found evidence of an extensively open landscape, including snail species – such as the rather smart and stripy heath snail *Helicella itala* – which are still found in suitable habitat in the local area.

At Ranscombe Farm Reserve, some lynchets – ridges across a hillside which form as repeated ploughing moves soil downslope to form a narrow, level terrace – have been tentatively ascribed as dating to the Bronze Age. Certainly, given the evidence from the Medway Gap, Ranscombe must have been farmed in the Iron Age, and this is backed up by finds of Iron Age coins at Ranscombe in recent years. A few years back, Plantlife was approached by an experienced metal detectorist called Tim who was looking for somewhere local to do a bit of detecting work, and he was given permission to explore the less sensitive parts of the reserve. As a result, we've been able to fill in a little more of the blank space in Ranscombe Farm's history. Finds have so far included two coins dating to the Iron Age, both different in design but both cast from a copper alloy known as potin.

Iron Age coin of the Cantiaci tribe, from the 1st Century BC, found at Ranscombe Farm. The face of the coin (left) shows a heavily stylised head of Apollo, while on the reverse (right) is a stylised figure of a bull. Photograph © Kent County Council and used under a Creative Commons licence CC BY 2.0 (https://creativecommons.org/licenses/by/2.0/)

One is dateable to the First Century BC, so it had been lying in the ground for more than 2,000 years. It measures about 2cm across and bears strange, stylized images on each side: on one face is a set of circles and lines, while on the other are some curved ridges which are vaguely reminiscent of the famous chalk carving of the Uffington White Horse in Oxfordshire. Both designs look almost completely abstract, but it turns out that the Iron Age people who made the coins had based the design on copies of copies of certain continental coins which had originally minted long before in the Mediterranean port city of Marseilles – itself originally a Greek Settlement. The original coins had a design on one side showing the head of the god Apollo, and on the other the image of a bull, but with each copy, the design of the coin had been changed little-by-little, like the game of Whispers but in solid form, until it became just a stylised series of marks that hints very strongly at a process of early financial trade and exchange.

If farming started in the Fertile Crescent 10,000 years ago, and got to Britain a little less than 6,000 years ago – and if Ranscombe Farm may have been an arable farm some 2,000 years ago in the Iron Age – then when did the site's cornfield weeds arrive? Fortunately, there is some fossil and archaeological evidence that helps us answer that question. When Mesolithic people were living at Ranscombe, it turns out that some two dozen species were already present in Britain which would later become regular cornfield weeds. These were plants that had made their own way north and west through Europe: without human assistance, they had been able to follow the retreating ice sheets and make it across into Britain before it was cut off by the North Sea and English Channel. One of these plants was field madder *Sherardia arvensis*, low-growing, with whorls of pointed leaves and tiny, pink, four-petalled flowers. Another was the common weed fat-hen *Chenopodium album*, an early foraged food: the seeds and leaves can both be eaten. The attractive and delicate geranium, long-stalked crane's-bill *Geranium columbinum*, and the current-day rarity ground-pine, were also among early residents. Presumably these annuals hung on where soils were dry and shallow and easily disturbed, or were able to respond to poaching of the ground by large herbivores, or grubbing by wild boar. Field madder certainly does well today in sparse, dry grassland and on sand-dunes, while ground-pine has a reputation for sudden reappearance following deep soil disturbance. In its heartland on the central North Downs, the short, grey-green tufts of ground-pine have a habitat of popping up on the ground where a new gate- or fence-post has been dug in, and have done just that at Ranscombe Farm on occasion. All these plants must presumably have been rare and scattered in the wild Mesolithic landscape, but the start of farming in the Neolithic, and its expansion through the Bronze Age that followed, must have provided them with a ready home

Field madder - a plant that became a cornfield weed, but which was already at home in Britain before farming arrived,

Ground-pine will pop up when grassland or scrub is disturbed, but it has also become a cornfield weed. Photograph: R I Moyse/Plantlife.

where they could easily persist, increase and spread.

At least another two-dozen species of our cornfield weeds arrived in Britain with the first farmers, hitching a ride amongst cereal seed. These included those wild plants which had become part of the farming field-scape in its early stages in Southern Europe and the Middle East, and which had been carried northwards and westwards by the first farmers as they moved across Europe. The archaeological record shows that by the end of the Bronze Age, around 2,800 years ago, a whole suite of new plants had arrived and settled into the farmed landscape, and these include what are now amongst our most important, beautiful and often rare cornfield weeds: stinking chamomile, opium poppy, rough poppy, prickly poppy and narrow-fruited cornsalad are all among this contingent. Perhaps surprisingly, common poppy and corncockle only start to show up in archaeological remains in Britain during the Iron Age, which in Britain was a period of around 850 years from the end of the Bronze Age to the start of the Roman occupation in 43AD. Why they should be so late to the party is not clear, but their arrival at this time may reflect trade in cereals between the Iron Age Celtic people of Britain and their counterparts in Europe, or between the Britons and the Roman Empire in the period before the Roman Invasion.

After the Romans successfully invaded and conquered Britain in 43AD, they stayed for another 400 years. Their coins have also been found at Ranscombe Farm – unsurprising given that Roman Rochester was less than an hour's walk away and that the first Roman arterial road in Britain – later called Watling Street and nowadays the A2 – lay less than a mile to the north. Some cornfield weeds which are common and very widespread today make their first appearance in the archaeological record in Roman and Romano-British times: these include red dead-nettle *Lamium purpureum* and henbit dead-nettle *Lamium amplexicaule* – two members of the mint and dead-nettle family with purple-red, two-lipped flowers – as well as field forget-me-not *Myosotis arvensis* with its tiny, five-petalled flowers of baby-blue. Again, it may be trade in grain between Roman settlements in Britain and mainland Europe that led to these species' eventual arrival.

Red dead-nettle first arrived in Britain in Romano-British times

By late Medieval times at the end of the 1400s, the plants that make up Ranscombe Farm's cornfield weed flora would have all been in place, except for a small handful of later arrivals. Maps from the mid-17th Century suggest that the field pattern seen at Ranscombe Farm Reserve today was probably already established by the end of the Medieval period, and the fact that most of the woodland at Ranscombe is regarded as ancient – i.e. that it was already there by 1600AD – backs this up. The lay-out of the site would certainly have been recognisable to modern eyes, perhaps even down to some of the tracks and footpaths, though not everything would look quite the same. Elm trees *Ulmus* spp. would have been a significant feature of the landscape, this being a period before Dutch elm disease *Ophiostoma* sp., and their leafy branches may have been cropped to provide forage for livestock (English elm *Ulmus procera* is still present at Ranscombe, but one of the effects of Dutch elm disease is that the trees generally remain small). There may have been one or two additional hedgerows, and some tracks may have shifted since then: for instance, where the current North Downs Way crosses the valley in the southern part of the site, the track once took a longer, more shallowly-sloping route that would have been easier for a horse-and-cart than the current path, which

A Medieval silver penny of Edward I (1272-1307) or Edward II (1307-1327), discovered at Ranscombe Farm Reserve. Photograph © Kent County Council and used under a Creative Commons licence CC BY 2.0 (https://creativecommons.org/licenses/by/2.0/)

goes straight down and up the valley sides and is a steep, slow climb even on foot. Perhaps surprisingly, Medieval Ranscombe would also have been less wooded than at present, particularly along the southern boundary where much of the woodland post-dates the coming of the railways in the 19th Century (and which will be discussed in a later chapter).

One of the bigger differences between modern day and Medieval Ranscombe would have been the number of people living on and working the land. Evidence of these people is sparse, but we do have hints. One of the first metal-detector finds at Ranscombe was a very handsome silver penny which dates from the first few years of the 14th Century; the wording on the coin shows that it was minted at Canterbury, just 30 miles away along the old Roman road. It may have represented a day's wages for the unlucky person that lost it, though perhaps it didn't just drop through a hole in someone's pocket, but was being safely stored in a house that was later abandoned and has long-since vanished. It could be that there were a number of small dwellings scattered across Ranscombe in Medieval times – certainly there are signs of buildings where none now stand and where none are indicated on the earliest maps. Dave May has pointed out that the very oldest maps not only show twelve buildings around the settlement or farm of Ranscombe (where the private houses are today), but that there are references in old documents to a number of crofts (a small building with 4 acres – about 1.6 hectares – of land to support a single family) in and around the site.

In 2007, what turned out to probably be a well was discovered on the edge of Kitchen Field after a tractor wheel dropped into soft ground. Excavation by Kent Underground Research Group found the upper parts of the shaft were lined with flint, while fragments of Medieval pottery, dating up to but not beyond the 14th Century, were found amongst the excavated spoil. The well is three-quarters of a mile from the former Ranscombe Farm buildings, so must have been dug to supply a much closer dwelling. Perhaps it was linked to the (perhaps 13th Century) windmill which seems likely to have stood on the ridge which is still called Mill Hill today – again, this building is not shown on maps but is indicated by some fragments of possible foundation masonry – though this seems unlikely as this would have put the mill more than a third of a mile from the well. So there must have been some vanished building, and perhaps more than one, at the western end of the reserve, because water is much too heavy to carry far (and Dave May suggests that the name Kitchen Field may even be a corruption of the Old English 'cyte' meaning hut). Elsewhere on the reserve, tucked into the edges of woodland, are some level, apparently deliberately flattened areas which Dave believes may have once been the locations of simple, small dwellings. One of these, which is tucked into what otherwise is ancient, mature woodland, shows signs of having been deliberately limed with chalk to sweeten the heavy soil, and still grows a dense crop of nettles – a good sign of past enrichment by manure.

Botanists in Britain (and, indeed, Europe) have a bit of a focus on the end of Medieval period in the way they think about the origins of different plant species. This is because, in order to understand how plants have come to be where they are, botanists have put all the wild plants in Britain into one of three categories. There are 'native' plants, which are those thought to have arrived in Britain without any help (whether deliberate or accidental) from people. These 'natives' include all the plants which survived in Britain during the last Ice Age, or those which colonised from mainland Europe before the North Sea and English Channel were finally completely flooded during the Mesolithic period. But it might also include species which blew in as seed in more recent times – though not any that arrived in the mud on vehicle tyres, which can make for long and inconclusive debates about how a newly discovered plant species may have arrived. Reaching a decision may be especially difficult if, as in the case of at least one discovery of an orchid species new to Britain, it is a plant with tiny, wind-blown seeds and is discovered close to a major road to a coastal port, so that either route is entirely feasible.

Those plants which arrived in Britain because people brought them here, whether they meant to or not, are either classified as 'archaeophytes' if they arrived before 1500AD, or as 'neophytes' if they arrived after this time. This date roughly coincides with what historians see as the end of the Medieval period and the start of the Early Modern period, when there were major changes in society: the long period of civil war known as the War of the Roses had recently come to an end; movable type had recently been invented, revolutionising the production of books; the Reformation was about to start in Europe; and major advances in science were about to be made. Critically for botany, Columbus had just reached the Americas, opening a whole new source of plant imports to Europe.

Although the cut-off date of 1500 is useful rather than reflecting some kind of objective, scientific truth, categorising plants as native, archaeophyte or neophyte is useful if you are trying to understand why plants are where they are, or why they behave as they do. For example, we know that most cornfield weeds are largely or wholly restricted to arable farmland, and rarely or never found in natural habitats: they never had a place in wild Britain before the arrival of farming, and the habitats where they do grow as truly wild plants are thousands of miles away. Nonetheless, these plants found their niche in Britain long ago, and they have had the time to spread and occupy this niche wherever it occurs. As a result, their national distribution is pretty settled: they've already had the chance to explore the length and breadth of these islands, and to decide where they want to live. By contrast, more recent arrivals may be expected to still show periods of expansion and change in their distribution, simply because they haven't been here long enough to have found all the places that might hold them.

But where classification into native/archaeophyte/neophyte is less useful is when we give value to a particular plant or group of plants and decide which are worthy of conservation and which not. Here, there tends to be a split between natives and archaeophytes on the one hand (with archaeophytes being treated as 'honorary natives') and neophytes on the other, so that when plants are Red-listed to determine how threatened they are (and, by extension, how much conservation effort ought to be directed their way), neophytes are excluded from the process. It is entirely understandable that people might want to conserve plants that have been here so long that we've forgotten when they first arrived and have had to go to archaeologists to tell us. Our cornfield weed flora is so familiar to us today, so much a part of how we see the countryside, that only a brave or peculiarly contrary person would argue that they are not important. But why not value the more recent arrivals? After all, all those archaeophytes were newly arrived at one time: if some kind of Iron Age conservationist had decided they all had to go because they were a threat to native plants, would we be thanking them now?

Nowadays, we hear a lot about invasive, non-native plants: those species which have been recently introduced to Britain and have spread so rapidly and are so competitive that they are driving out our other wild flora. It is true that some of these neophytes are a serious and costly problem: amongst the better known of these are Japanese knotweed *Reynoutria japonica* and rhododendron *Rhododendron* spp., both of which occur (in mercifully small amounts) at Ranscombe. But the proportion of neophytes which are actually or potentially damaging to other wildlife in Britain is tiny. There are maybe 1,300 neophytes growing wild in Britain – perhaps more – yet Defra lists only 36 as being of special concern. Admittedly this looks like a serious under-estimate of the real number of problem species – it excludes such species as New Zealand pigmyweed *Crassula helmsii*, which seems impossible to keep out of any ponds or wetlands in England – and the number of invasive non-native plant species included in the official Biodiversity Indicators for England is larger at 80. Nonetheless, even this larger figure is just six percent of the total number of recorded neophytes. Incidentally, the list of 80 problem species includes a few plants beloved of wildlife gardeners. One such is butterfly-bush *Buddleia davidii*, which can be tremendously invasive and which has to be kept under close control in some parts of Ranscombe Farm Reserve.

It is also the case that many native species can be invasive and pose real problems for the conservation of wild plants and other wildlife. A recent article in the newsletter of the BSBI gave a rough-and-ready assessment of the most invasive plants in Dorset and Cornwall, and found that five of the top ten in each county were natives. These included plants like common ivy *Hedera helix*, bracken *Pteridium aquilinum* and gorse *Ulex europaeus*, all of which are showing rapid spread in wild habitats, much of which is likely to be due to changes in the way land is managed.

So, on this basis, shouldn't we care a little more about the fate of more recently-arrived plants and take their conservation more into account? At the very least, shouldn't we regard them as interesting and potentially valuable additions to our flora? Well in excess of 400 plant species (excluding mosses and liverworts) have been recorded so far at Ranscombe Farm Reserve, and the total increases every year. Of these, just three-quarters are considered native to Britain, while around one plant in every in six is an archaeophyte, and about one in twelve is a neophyte. Sometimes, as I have said, it might be necessary to deal with problems caused by some invasive neophyte plants – trying to eliminate Japanese knotweed, for example, or controlling butterfly-bush where it is invasive on chalky soils – but if Cape-gooseberry *Physalis peruviana*, an annual in the nightshade family and native to South America, turns up on disturbed ground on the edge of an arable field, or pale yellow-eyed-grass *Sisyrinchium striatum* (a member of the iris family) turns up in the woods (and these are both genuine examples from Ranscombe), this just adds to the interest of the site. Not least, you wonder how the plants got there, and what this says about the ways in which plants are able to move through the landscape.

In botanical and conservation terms, these 'non-natives' which have found their own way into the site can be more interesting and a great deal less problematic than deliberately but inappropriately introduced native plants. Plants which have recently arrived in Britain would be expected to turn up in odd places, as it takes some time for their distribution to settle. But if a native plant or an archaeophyte turns up somewhere unexpected, it can be hard to know whether the occurrence was natural or the result of deliberate introduction – especially in the case of showy and attractive species. This just confuses the picture of the plant's behaviour and natural distribution in the landscape, and, in the case of rare species, this may just make their conservation more difficult. As an example, meadow crane's-bill *Geranium pratense* may owe its presence at Ranscombe Farm Reserve to the sowing of wildflower seed on the two land-bridges across the Channel Tunnel Rail Link that was constructed in the late 1990s. Or at least that's my assumption, given that meadow crane's-bill is only sparsely distributed in Kent. But I could be wrong, and the problem is that I just don't know for sure: it's not so much that it matters whether the plant was deliberately introduced or not, rather that having an answer – one way or the other – would help us understand the plant better and actually give more meaning to its presence.

Some deliberately introduced native plants are easier to guess. Scots pine *Pinus sylvestris*, for example, must have been planted many years back, as it's highly unlikely that it has remained present on the Kent Downs – so far outside its known native range in the Scottish Highlands – since its first colonisation following the last Ice Age. But it can be harder when the plant introduces itself by escaping from a nearby garden. Tutsan *Hypericum androsaemum*, for example, is a shrubby plant related to the St John's-worts, and its attractive, five-petalled, yellow flowers mean that it is a popular garden plant as well as being a widespread British wildflower. The flowers produce black berries which are eaten by birds, and so plants are easily spread around the landscape. Which means that it is impossible to know whether the presence of tutsan in Cobham Woods is due to its being a long-standing member of the woodland flora here, or whether it has at some stage been carried in from a local garden as a seed in the guts of some bird. Though it's just as difficult to know whether this really matters very much anyway: it got here in its own way and without anyone deliberately helping it, so all power to it.

Probably the one plant at Ranscombe that really tests whether the distinction between natives, archaeophytes and neophytes is really important for conservation is our old favourite, rough mallow. At Ranscombe Farm, because of its rarity in Britain and its history at the site, it is one of the plants that has received most attention in terms of work to try and increase its numbers. As a rare plant, it has previously been Red-listed as being at threat of extinction, and it was on the Red List when Plantlife was working to conserve it at Ranscombe Farm in the 1990s. However, since then, the scientific consensus has shifted and it is now thought that rough mallow was probably not present in the wild in Britain before 1500: this means that it is now considered a neophyte, and as a neophyte it is not included in the Red List. Nothing else about it has changed. It hasn't suddenly become more common. All that has happened is that a judgement has been made that rough mallow was probably brought to Britain from Europe, was grown in gardens, and then escaped into the wild sometime after 1500. Though, conversely, some authors have suggested that rough mallow may possibly be native at Ranscombe Farm, even if not elsewhere, though on what basis is not entirely clear. Whatever its status, the plant's long pedigree at Ranscombe Farm – and, not least, the people it has met – surely means it is of historical and even cultural interest.

Rough mallow is an interesting plant which seems to be able to grow as an annual or as a biennial: in either case, it dies after flowering and has to grow again from seed (and the seed can last for many years in the soil, in common with most cornfield weeds). Germination of the seed does seem to rely on the soil being turned over, so it occurs as a cornfield weed in the margins of three different arable fields at Ranscombe, mainly on chalky soil. Where rough mallow is forced to grow as an annual by the yearly cultivation of the arable fields, the resulting plants tend to be small; sometimes, on the poorest soils, they can be just a few inches tall with one or two flowers. But if the cultivation is missed for a year (and this can happen if spring cultivation in one year is followed by autumn cultivation the next – or if farm machinery breaks down at an inopportune time), then plants can behave as biennials. In this case, the seeds germinate in spring or summer of the first year, but the resulting seedlings pass the winter as young plants and don't produce flowers till the following summer. Having had longer to grow, these biennial plants are generally much larger and have lots more flowers. What I find most interesting, though, is that rough mallow looks most at home, not in the arable fields, but in grassland or on

Rough mallow - once considered an ancient introduction, now considered to have been introduced after 1500AD and consequently losing its previous official conservation status. Photograph R I Moyse/Plantlife.

the edges of woodland – though only where some pretty heavy soil disturbance has taken place and brought buried seed up to the surface. In such places the plants can get much larger, especially given plenty of nutrients: one plant which grew on the edge of the site of an old bonfire grew to several feet across, with a series of stems radiating out across the ground like the spokes of a cartwheel, and bore more than 300 flowers. Rough mallow plants which find themselves amongst tall, grassy vegetation don't end up getting shaded out like some cornfield weeds would, but can straggle their way up towards the light, apparently supported amongst the other plant stems by the stiff, bristly hairs on their stems and leaves.

Plantlife's early work for rough mallow at Ranscombe Farm also suggested that the plant acted much like a biennial, needing around 18 months between germination and flowering. My own experience suggests that it is a plant that will germinate almost any time from spring to autumn should the conditions be right (presumably the ground needs to be warm enough and with the right amount of moisture), and that it can flower in the same year if it has a long-enough growing period. It is the seeds that germinate too late in the season to flower that will over-winter and flower the following year, and so make the bigger plants. Interestingly, it seems to share this feature with ground-pine, which has had lots of mentions so far, but which I have not really described. This very rare member of the dead-nettle family is a distinctive little plant which gets its name because its narrowly-divided leaves and resiny scent give it the superficial appearance of a pine tree seedling, though this illusion is dispelled once the yellow, two-lipped flowers appear. Ground-pine has a very restricted distribution in Britain, with its heartland being the North Downs from Surrey to Mid-Kent and with something of a hot-spot around the Medway Gap, so it is one of the plants that people come to Ranscombe Farm Reserve to look for. Like rough mallow, it will use arable field margins, but it seems really to be a plant of chalky grassland, where its strategy is to lie in the soil as seed until something happens to turn the ground over and bring the buried seed to the surface. Germination of ground-pine tends to be in autumn, but it is possible this is just because that is when the ground is moist and is still warm from late summer sunshine. My evidence for thinking this is the experience of one very wet summer a few years back, when the number of ground-pine plants in Kitchen Field just kept increasing right through August as more and more plants germinated.

If rough mallow is not native at Ranscombe Farm Reserve, and is either an archaeophyte or a neophyte, then it found its way there from the species' heartland in Southern Europe. Whatever its history, it managed to find a home in the wild in Britain, and it has shown a remarkable persistence in surviving at Ranscombe Farm for more than two centuries. Most of Ranscombe Farm's cornfield weeds started out in the earliest farmed fields of the Eastern Mediterranean and Middle East, and they've been growing alongside us and our ancestors for 10,000 years and maybe for as long as 2,000 years just at Ranscombe Farm – now, that is persistence. And these are wild plants, even the archaeophytes that arrived and thrived alongside farming, even the neophytes that may have started off in a British garden. Because, although these plants were and are dependent on human activity for their presence here or even for their continued survival, they really persisted in spite of human endeavour as much as they did because of it. Even corncockle, which needed repeated sowing of its seeds alongside cereal grains, was, in effect, relying on its own wits to outmanoeuvre the human managers of its habitat. There is something to admire in that, something to celebrate, and if I have a bit of a wobble about finding plants like meadow crane's-bill at Ranscombe Farm Reserve, and suspecting that it was part of recent, deliberate sowing, then this may be why. Cornfield weeds are amongst the easiest plants to sow and grow – seed for many of the brightest, showiest species is easily bought, and you don't have to google far to see dozens of photos of obviously sown parks, road verges and field margins, all glowing with common poppies, cornflowers and corn marigolds *Glebionis segetum*. It may say 'wildflowers' on the seed packet, but plants like this are not wild but tamed. They are not living the life they were designed for and which they've carved out for themselves; instead they are living the life we have decided for them. They may look beautiful (and there's nothing wrong with that in itself) but the best and most important part of them – how they live, what they do, how they behave – is lost.

Chapter 5 - Into the trees

Arable farming and the cornfield weeds that go with it have a long history, but the history of woodland goes back a lot, lot further. The very first Mesolithic people to find their way to Britain some 12,000 years ago may well have found an environment that resembled today's cold, open, Arctic tundra, but 6,000 years later at the end of the Mesolithic, the rapidly warming climate meant that Britain was covered in woodland. Agriculture in Britain may go back 6,000 years, but woods and woodland plants were already established and dominant when farming got here. Ranscombe Farm Reserve today is a well-wooded site. Although it is perhaps best known and certainly most important for its cornfield weeds, only a relatively small portion of the reserve is managed for these plants. At the time of writing, about 30 hectares (74 acres) or so are dedicated to cornfield weeds, of which perhaps 10 hectares (25 acres) are especially rich in species, and the rest a work currently in progress. Slightly more of the reserve – around 31 hectares (76 acres) – is managed as grassland for wildflowers, though in this case only a tiny part, maybe half a hectare, is really rich in wild plant species, with the remainder being grassland in various stages of restoration from former arable. The area of woodland, though, is massive. At more than 130 hectares (320 acres), woodland makes up more than half the reserve's total area. It dominates the slopes in the northern half of the site, covering the sands and gravels on the ridge-top, rolling down the gravelly soils of the steeper slopes, and drifting over the shallower slopes where the soil has developed over the heavy, soggy clay-with-flints.

My guess is that woodland is probably most people's favourite kind of wildlife habitat. The common poppy may be the one plant which strikes the loudest chord in the hearts and minds of people in this country, being so tightly tied to the horrors of the First World War and the lasting impact this has had on national life. But take a poll on our favourite wildflower, and bluebell would undoubtedly top the list. Trees, woodland and wildlife are so intertwined in our heads that nature conservation and tree-planting are often seen as being more-or-less the same thing, despite this country having a lot more good, species-rich woodland than it does, say, flower-rich grassland. The reason for this must surely lie in the apparent unchangingness of woodland: the trees were there before we were born, and they'll long outlive us. This means that woodlands give a sense of stability and permanence to the environment, and they can act as a mental and emotional anchor in a world where change is increasingly rapid and frequent.

A large apple tree *Malus domestica* by the Town Road in Cobham Woods. This tree is growing wild, but has cultivated origins.
Photograph: R I Moyse/ Plantlife.

Almost all the woodland at Ranscombe Farm Reserve is regarded as 'ancient', meaning that it's been there since at least 1600AD, and probably for a lot longer. As with a lot of the remaining woods on the North Downs, Ranscombe's woodland sits mainly on the soils that are hardest to cultivate for crop-growing: the gravelly soils that are dry and lacking in nutrients, the wet and heavy clays, the steeper slopes. Some woodland does survive on soil that might otherwise have been put to the plough: Merrals Shaw, for instance, the little block of woodland on the flat plateau above the main entrance to the reserve, lies on clay-with-flints, soils which are fertile enough to have long been worth working as arable. It may be, though, that this woodland was saved by the wear-and-tear suffered by farm machinery in soil with such a huge density of flints at the soil surface (next to Merrals Shaw is the field called Prebbles on the old 19th Century tithe maps, but which the farmer knows as 'Brighton Beach' because of the huge number of surface stones). Ranscombe Farm does have some woodland on the lighter chalk soils, in spite of this being the easiest and most useful soil to work for arable – however, a good deal of this woodland (as discussed below) turns out to be of fairly recent origin.

Maps tell us that Ranscombe's woodland dates back to at least 1600, but it is hard to say how much further back in time its origins might actually lie. The received wisdom used to be that almost all of the North Downs was cleared of trees by Neolithic farmers, and that most or all of the woodland around today arose when Neolithic crop fields and grazing land were later abandoned. But wet clay soils and stony ground have always been hard to cultivate, so why would Stone Age, Iron Age or even Medieval farmers have bothered? Wood was, after all, a hugely important commodity until well into the Industrial Revolution – not just as the main structural building material, but as the fuel for cooking and heating – so to have some woodland would have been economically essential. Leaving trees to grow on the worst agricultural land would always have made sense.

This wouldn't mean, though, that the woodland at Ranscombe Farm today is the same as the woodland of Prehistoric or Medieval times. For instance, it used to be common to turn out livestock into woodland – subsistence farming means wasting nothing, and cattle might as well have a chew of the woodland vegetation – and maybe those bits of the site with gravelly or sandy soils would have had a more open, heathy vegetation because of this. But – who knows? – perhaps there is some chance that at least a bit of the woodland at Ranscombe Farm dates right back to the time the first trees colonised Britain at the end of the last glacial period. Just perhaps. But even if this is the case, the woods have been messed around with a lot since then, and the trees most of all.

Ranscombe Farm Reserve's woodlands look the way they do today because of the big fashion for planting sweet chestnut *Castanea sativa* that really took off during the 19th Century. Chestnut timber is resistant to rot, and makes great fence posts which have a long life standing in the ground and so do not need to be replaced so frequently. Sweet chestnut is also very fast-growing as a coppice tree: cut to the ground in the winter, the stump sends up a number of shoots which are taller than you or me by the end of the summer, and able to provide useful timber after just ten to fifteen years. The shoots from the coppiced stumps grow up straight, which is not just good for making fence-posts, but made the tree a very useful source of the many tall poles needed for growing hops to supply the brewing industry. Over much of the Kent Downs, as elsewhere, other tree species were grubbed out of the existing woodlands and replaced with sweet chestnut, and this happened to almost all the existing woodland at Ranscombe Farm. Amongst the sweet chestnut, which is virtually a monoculture in places at Ranscombe, a smaller number of larger trees, such as oaks *Quercus* spp., would have been allowed to grow big and tall to supply large timbers for construction – the nearby Naval Dockyard at Chatham used a lot of timber. The monotony was also broken up by bits of 19th Century ornamental planting, which at Ranscombe Farm included hornbeam *Carpinus betulus* and beech. There were even some planted conifers at this time, though more were planted during the 20th Century.

All this replanting means that if you go through the woods looking up, you can get an idea of the history of the woodland as part of the human landscape and local economy, but learn less about the land beneath or the wildlife it supports. There are some clues: where you see ash *Fraxinus excelsior* and field maple, it tells you that you are on more alkaline soils, and if you look close enough at the oak trees, you can spot where pedunculate oak is replaced by sessile oak *Quercus petraea* on the sandier and more acidic soils (though, in truth, many of the trees that generally pass as sessile oaks at Ranscombe Farm seem to be carrying some pedunculate oak genes: unsurprising, as these two species happily hybridise and the hybrids also hybridise, including with their parents, so that it's not at all easy to see where one species ends and the other begins). But, as important as the trees are to the woodland, if you really want to understand what's going on, you need to look down: the low-growing stuff in the woods is there because it wants to be, not because someone has put it there. It is true that there are many self-sown trees and shrubs in Ranscombe Farm's woodlands, but there are a lot more species of herbaceous plants growing beneath them on the woodland floor. Greater variety means a wider range of potential responses to variations in the woodland environment, and this means that it is the ground-floor plants that can tell you most about what's really going on beneath your feet.

We tend to think of plants as being in communities – forming 'habitats' – so that we talk of grassland plants, heathland plants, wetland plants, and so on, as if the different species in each of these communities all wanted the exactly same thing. But in practice every plant species is different, and experiences and responds to its circumstances in its own way. Plants are pretty smart and know a lot about what's going on around them; they can tell which way is up, where the light is, whether there's enough water, whether some insect or other is chewing on them and a lot more. Roots are able to react to soil chemistry or soil oxygen levels (roots need to breathe, too). Seeds have some mechanism to tell them whether it's a good time to germinate. Adult plants must have systems to tell them that it's a good time to flower (and that they have the spare resources available to invest in making flowers), or that it's better to put effort into vegetative growth, or even that things are so bad that it's time to give up growing for a bit and just hunker down till conditions improve. Different species of plant will respond to each of these things in a different way, so that, depending on the species, the same spot could be too dry, or too wet, or too alkaline, or too low in nutrients, or too humid, or too sunny – or just right. The woods at Ranscombe Farm Reserve are a great place to watch all this in action, and especially in the northern half of the reserve at Cobham Woods. Here the underlying geology varies so much across the woodland that different parts of it may have soils which are wet or dry, acid or alkaline, rich in nutrients or dirt-poor. This makes for conditions which suit a lot of different plants with a lot of different lifestyles and behaviours, and provides the curious with lots of opportunities to stare at a particular plant and wonder why it is there and what it is up to.

At the highest point in Cobham Woods, where tracks and paths have worn away the surface soil, you can see the underlying sands and gravels. There is an abundance of small, dark, round pebbles which look as if they were recently put here to surface the paths, but which were actually dropped into a shallow sea anything from 5,000,000 to 23,000,000 years ago. But even if you couldn't see what the soil looked like, the plants would tell you that it's dry and acidic and heath-like. Where there are gaps in the tree-cover, you find the brilliant yellow flowers of broom *Cytisus scoparius*, the smart purple towers of flowering foxgloves *Digitalis purpurea*, and sometimes the scruffy but no less attractive heath groundsel *Senecio sylvaticus*. These are not plants that you will find growing on chalky, alkaline soil. There are other plants here, though, that are fussy

Common cudweed growing on gravelly soils in Cobham Woods - elsewhere on the reserve, it grows on chalk. Photograph: R I Moyse/Plantlife.

about having good drainage, but less bothered about soil chemistry: as both sandy and chalky soils are generally very dry, with any rain running straight through, such plants can be found in both acidic and alkaline spots. At Ranscombe, one of these is common cudweed *Filago germanica*: this annual will pop up in the woods where the gravelly soils have been disturbed or turned over, but also turns up (often alongside the very similar broad-leaved cudweed) in the chalky, alkaline but equally dry arable fields.

Then there are those plants which are fussy enough to want the soil not just to be acidic, but to also be both moist and well-drained at the same time: they want plenty of water, but they don't want it sitting around too long. Trailing St John's-wort *Hypericum humifusum* and trailing tormentil *Potentilla anglica* are unrelated, but both have creeping stems and yellow flowers, and both find a home in Cobham Woods in the remains of an old pond: though sat over gravel, and normally drying up in the summer, the pond still holds a bit of water after periods of rain. The two plants are both low-growing and attractive, with the petite looks and habits of garden alpines: trailing St John's-wort has the star-shaped flowers of its common

The yellow flower and trailing stems of trailing tormentil, growing on the edge of a ride in Cobham Woods.

and more widespread relatives, and trailing tormentil has four or five broad petals giving it almost a Tudor rose shape (and hinting that it is, indeed, a member of the rose family).

For me, trailing St John's-wort is a bit of a favourite: it is not a plant you see every day, and it has a neat, small-but-perfectly-formed appearance, so that I find I'm admiring it in the same way I might admire a well-made scale-model. On the other hand, trailing tormentil is one of those plants that invariably present a struggle. It is a close relative of both the heathland plant, tormentil *Potentilla erecta*, and the common grassland flower, creeping cinquefoil *Potentilla reptans*, and can hybridise with either of these species. To decide whether what you're looking at is a hybrid or is actually the real trailing tormentil, you need to see fruit: the hybrids are almost completely sterile, so only the real trailing tormentil has a fruit with fully-formed seeds. Which means revisiting the plants through the flowering period until you finally find that fruits have (or haven't) formed. Some plants are annoying in this way – they demand your attention and make you keep coming back until you can finally identify them.

From the gravelly high-point of Cobham Woods, the track called the Town Road heads eastward and downhill, and the soil looks deeper and darker as you move from the pure gravels onto what geologists call 'head' – a mix of sand, clay, silt and gravel that formed on and then slid down the slope under the influence of millions of years of rain and snow and ice. Here

Deliberately created and managed woodland open space, like this ride in Cobham Woods, provides a habitat for many woodland plants and animals. Photograph: R I Moyse/Plantlife.

are stands of autumn hawkweed *Hieracium sabaudum*, which open its dandelion-like flowers in late summer, and which shows that the banks each side of the track remain dry and well-drained. Further down the slope, shade, wetter soil and more humidity seem to suit lily-of-the-valley *Convallaria majalis*, which makes a large patch close to the path, but can only be coaxed into flower where gaps in the canopy let in a little more light. From here, you can keep going downhill by turning right just before the High Speed railway, and you soon see why the old name for this particular area is Clay Pond Wood: the impervious soil keeps puddles wet for weeks (in some years, you'll find tadpoles in them), and is perfect for damp-loving plants like wood speedwell *Veronica montana* and marsh thistle *Cirsium palustre*, even for wetland specialists like brooklime *Veronica beccabunga* and square-stalked St John's-wort *Hypericum tetrapterum*. Beneath and giving rise to the soil here is the clay-with-flints which formed sometime in the last 20 million years or so as the surface chalk was weathered by successive periods of glaciation.

You would think that wet soil would dry out pretty quickly in woodland, given that trees are very effective at sucking water from the ground through their roots and sending it out through their leaves into the atmosphere: trees can and often do dry out and damage wetland habitats. But it is also the case

Lily-of-the-valley will grow in shade, but only flowers in the light. Photograph: R I Moyse/Plantlife.

that their shade and shelter can keep the soil-surface damp, prevent puddles from drying out, and favour plants that might find it too dry out in the open. This is one of the reasons why open space within a woodland can be so different to open space outside the woodland, and why the creation and maintenance of gaps in woodland is so important for wildlife. Such gaps in woodland are certainly important for many of the plants mentioned above, from trailing St John's-wort, to lily-of-the-valley to wood speedwell, all of which need to be both in woodland and in the light. At Ranscombe Farm Reserve, a lot of work has gone into opening up small, permanent open spaces (generally known as glades) and creating long openings (or rides) along tracks and paths. This might seem perverse, given that there's plenty of unwooded ground available at Ranscombe, and loads more in the surrounding landscape. It is also an expensive and time-consuming process: as well as the initial labour needed to open up a gap wide enough to let in sunlight, there's the ongoing work to maintain a central grassy area to the ride or glade, and to keep the edges low, dense and scrubby. But the result is worth it, as open space within a woodland supports a different range of plants and animals to the open space outside woodland (though they do overlap), and this includes plants which are very much woodland species but still need more sunlight than they would get under a closed tree canopy.

Think of the blue spikes of bugle *Ajuga reptans*, such a favourite of spring-flying bees and butterflies, or primroses *Primula vulgaris*. Woodland open space is still part of the woodland, even if there are no trees overhead.

A very particular case in point at Ranscombe is hairy lady's-mantle *Alchemilla filicaulis* subspecies *vestita*, which is a clumsy and rather plain name for an attractive little plant. The lady's-mantles are a group of plants in the rose family, and are unusual in that many species – in fact all the British species – produce seed without the flowers being fertilised, so that each new seedling is a clone of its parent. This means that small genetic differences persist from generation to generation, producing a range of species which are sometimes only very slightly, but always consistently, different. Perhaps the best-known species to most people is soft lady's-mantle *Alchemilla mollis*, a garden plant which produces mounds of gentle, green foliage topped with frothy heads of golden-green flowers, and which sometimes escapes into the wild. Most of Britain's native lady's-mantles like the weather cool and wet, and so are far more common in the north and west of the country. Still, some three species did once occur in Kent, though two have now disappeared from the county, leaving just one species, hairy lady's-mantle, at just one last location, this being a single woodland track in Ranscombe Farm Reserve. Here, it benefits from the light that comes through the deliberately maintained gap in the trees, but also needs the damp soil which comes with the woodland's shelter, humidity and dappled shade. Nationally, hairy lady's-mantle is not rare, and in Northern and Western Britain it is widespread in rough pasture on mountainsides, on wood edges and on road verges. At Ranscombe Farm it still manages to hang on, but it struggles: although it's normally a perennial, the adult plants at Ranscombe seem unable to persist for any length of time and behave very much as if they were annuals, with the tiny population consisting of short-lived plants and undoubtedly depending on regular production of seed. The fact that Kent's hairy lady's-mantle is making its last stand in a small, sheltered piece of open space in a woodland emphasises just how important gaps in woodland can be. Though it does look as if Kent is just too hot and dry for these cold weather plants, and, with the climate changing so rapidly, it's really not clear whether any amount of conservation work will be able to save the county's last remaining wild lady's-mantles.

So far, I've concentrated on the plants in the woodland open spaces, because they are many and varied and probably say most about what's happening below the ground in the geology that shapes the land-form and the soils. If the open-space plants help to draw the geological detail, the plants that prefer life under the tree canopy paint with a broader brush, as a lot of them are not especially fussy when it comes to the soil, so long as it's deep enough. However, they do have some preferences and, in any case, it

Hairy lady's-mantle in the semi-shade on a damp track in Cobham Woods, the last remaining site for the species in Kent. Photograph: R I Moyse/Plantlife.

is these plants that produce those big, broad sweeps of spring-time colour that let us know that the Earth is still chasing round the sun and that summer is coming. Bluebell, our all-time number one favourite, is widespread throughout the woodland at Ranscombe, though it does better on damper and more acidic soils. On chalkier soils, bluebells tend to lose out to the creeping army of the under-stated dog's mercury *Mercurialis perennis*, which can form a wide, bright-green carpet. On the richer clay soils you can find yellow archangel, which pushes up spikes of deep-yellow, two-lipped flowers, the delicate pink flowers of cuckooflower *Cardamine pratensis* and the brilliant, shining yellow flowers of lesser celandine *Ficaria verna*.

Bluebell in particular is a special plant on our little cluster of islands. It is now the case that picking flowers is generally frowned upon (though in truth a little picking of common flowers causes no harm at all), but people used to regularly and happily pick bluebells in spring. When I was a boy, we'd bring home a few handfuls to put in vases on the mantelpiece. Some people could be seen collecting armfuls – filling the car boot – though even this excess didn't dent populations. Now people across the country still go on bluebell walks for charity or for fun every spring. Google 'English woodland' and see how many pictures show a carpet of bluebells under a high canopy of tall trees.

It turns out there's a reason that bluebells are so prominent in the national consciousness. In a global context, bluebells are plants of the western fringe of Europe, their world distribution stretching from Britain and Ireland, into The Netherlands, and then down through Belgium and France into Spain and Portugal. They especially like the damp oceanic climate, and Britain and Ireland together are thought to hold more than half the world's bluebell population. The dampness of our misty little islands, and the mild, wet winds coming in across the Gulf Stream must account for much of this, though other factors may come into play, especially to account for the sheer abundance of bluebells in some woodlands. This abundance is often attributed to our long history of woodland management by coppicing, which boosts bluebelll growth by giving them a good dose of sunlight for two or three years out of every fifteen or twenty. But the 300-year absence of wild boars from Britain, after their eventual extermination in the 17th Century may, also have played a part, as wild boars are fond of rooting-up and eating bluebell bulbs.

If you live in Eastern and especially South-eastern England, it might actually come as a surprise to learn that bluebells are not so much a woodland plant as you might think. Particularly in Western Britain, they will persist happily in pastures and in coastal clifftop grassland. What matters to bluebells is damp soil and not too much scorching summer sun, and woodland is ideal for providing these conditions over much of the drier parts of Britain. Couple that with bluebell's particular lifestyle, which is to produce leaves and flowers in early spring, taking advantage of spring sunlight before the trees come into leaf, then to retreat back into a bulb for the rest of the year, and you have a perfect combination. Bluebells also like a bit of stability in the places they live. Though they happily reproduce by seed, the seeds are large and heavy and generally don't move very far. As a result, it is difficult for bluebells to move about the landscape and especially hard for them to find their way into isolated patches of suitable habitat. This means that in South-east Britain – though not in the damper North and West – their presence is a good indication of long-established or even ancient woodland.

Many people are familiar with the term 'ancient woodland', but perhaps not so many understand its full significance. Ancient woodland, in the strict sense, is defined as woodland which is considered to have been continuously wooded since at least 1600AD, with the detailed ancient woodland inventories produced by the Forestry Commission being based on information from old maps. The reason that ancient woodlands are so very important for the conservation of biodiversity is because, like bluebells in the drier parts of Britain, many specialist woodland species are slow-moving and find it extremely difficult to make their way into newly created patches of woodland. These are pipe-and-slippers, stay-at-home types for which jumping across unwooded ground is a virtual impossibility; and for many species it can take years to make any headway into new woodland even where this is immediately next-door and all they need to do is shuffle across. As a result, ancient woodlands are generally much richer in a wide range of specialist woodland species than any more recent woodland. Those plants which are only or primarily found in ancient woods are known as 'ancient woodland indicators', and can be used to identify woodland as ancient even where there is no information available from old maps. What counts as an ancient woodland indicator does, however, depend on where you are in Britain – for example, as I've mentioned, bluebell is an OK indicator in the drier parts of Eastern and South-east England, but it definitely does not work in the west, where the wetter weather means that bluebells are far from restricted to woodland. You cannot rely on just one indicator, in any case: find a good number of different indicator species, and you may be able to say that a woodland is very likely to be ancient. But always caveat your decision: wildlife has never read the rules and will inevitably surprise and embarrass you by doing something the books say it never does.

A quick tally-up gives a total of some 40 ancient woodland indicator plants that can be found in Ranscombe Farm Reserve, but some of them do illustrate the pitfalls of trying to guess the age of a woodland by its plants. Cobham Woods – the large block of woodland in the north of the reserve – is most definitely ancient woodland: this much is clear from old maps, as well as from a suite of indicator plants (plus the presence of the mighty ash-black slug *Limax cinereoniger* – which can be up to 20cm long – an ancient woodland indicator mollusc!). These indicator species include trees, such as sessile oak and wild cherry *Prunus avium*, as well as the peculiar, prickly, low-growing shrub, butcher's-broom *Ruscus aculeatus*. Others are grasses such as wood meadow-grass *Poa nemoralis* and wood melick *Melica uniflora*, or the grass-like wood sedge *Carex sylvatica* and southern wood-rush *Luzula forsteri*, in addition to some plants I've already mentioned such as wood speedwell, lily-of-the-valley and yellow archangel. These plants are probably fairly reliable indicators, but some other plants generally regarded as ancient woodland indicators may not be so reliable at Ranscombe Farm. Tutsan may have its origin in nearby gardens and have been brought in – and pooped out – by birds. The same may also be true of guelder-rose *Viburnum opulus*, which can be found here

An indicator of ancient woodland: southern wood-rush is common along the southern edge of Cobham Woods. Photograph: R I Moyse/Plantlife.

and there in Cobham Woods, but which has also popped up in grassland on the reserve. Many of the hornbeam trees at Ranscombe Farm are most definitely unreliable indicators – those along the Town Road in Cobham Woods were deliberately planted to create a landscaped avenue when the woods were part of the Cobham Hall estate. Though these particular hornbeam trees are valuable as old trees with a range of wildlife-friendly features including dead wood and rot-holes – and are happily reproducing – they are certainly not there because of the age of the woodland.

But it gets more confusing and interesting when supposed ancient woodland indicators turn up in woodland which is clearly not very old at all. On the south side of the reserve, above the village of Cuxton, is Mill Hill Wood. This wedge of woodland certainly looks old, with its massive beech trees and even a number of interesting ancient woodland indicator plants. But a map from the end of the 18th Century – barely more than two centuries ago – shows the area to have largely been unwooded at that time. All that is shown on the map is a few lines of trees, as well as some narrow strips of woodland which were then, as now, linked to an existing, small and clearly ancient block of woodland called Bishop's Wood (the name dates from the 13th Century, when the tithes payable on this block of woodland were awarded to the Bishop of Rochester after he took a local landowner to court). These wooded areas probably make up only about an eighth of the woodland existing is this area today. Presumably, this had been the state of affairs for some time: if Mill Hill was really the site of a long-lost Medieval windmill, then this would not have worked very efficiently tucked within the shelter of trees.

The 1842 tithe schedules, which listed the charges payable by occupiers of different areas of land, show the continued presence of the woodland which existed at Mill Hill at the end of the 1700s. But sometime during the few decades between these dates, an additional area equivalent to roughly half of the current Mill Hill Wood had also been planted with trees; the schedules list it as 'plantation', so it was not just land that had been abandoned and left to turn to scrub and woodland. Of the remaining area, more than a fifth of what is now woodland was arable farmland and a further fifth was pasture grazed by farm livestock.

In around 1860, the London, Chatham and Dover Railway Company built the railway line which now forms the southern boundary of Ranscombe Farm Reserve. The new line cut through Mill Hill, so that part of the plantation and pasture lay to the south of the new railway, and part to the north. An 1898 Ordnance Survey map shows that the arable land was still present towards the end of the 19th Century, though the pasture to the south of the railway had clearly been abandoned and was scrubbing over. By the end of the 1930s, the arable area had been abandoned and was largely tree covered, and by the 1960s, the whole area was wooded: just a single small patch of open grassland of about one-third of a hectare (just under an acre) was all that remained of the original 2.7 hectares (more than 6.6 acres) of pasture.

This means that, today, Mill Hill consists almost entirely of woodland less than 250 years old. Yet take the path that runs up through the wood and you can find a good number of ancient woodland indicators throughout this young woodland, many in places well away from the original ancient woodland areas like Bishop's Wood. There is the vividly blue-flowered wild columbine, the low-growing shrub spurge-laurel *Daphne laureola*, stinking iris *Iris foetidissima* with its sword-shaped leaves, even lady orchid in places. These are all supposed to be plants that struggle to find their way into new woodland, which means that if we can understand how they came to be at Mill Hill, then we may learn something about the ways in which ancient woodland plants are able to get a foot-hold in new places: this may be hugely important given how little ancient woodland is left.

The history of the wood may be the key here: not only has Mill Hill been through the changes which I describe above, but much of it was felled by the Great Storm of 1987, after which most of the fallen trees were removed and a lot of replanting carried out (though much of the replacement planting was quickly overtaken by self-sown trees). The heavy work of removing acres of fallen trees must surely have churned up great areas of ground, and it may have been this scuffing and dragging and tracking which served to move plants about.

Take dog's mercury, for instance. This quiet little perennial plant is without showy flowers, but its creeping underground stems mean that it can form a dense carpet in woodlands on chalky soils. Its progress, though, is slow and steady: one study of hedgerows of known ages showed that dog's mercury was creeping along at a speed of up to one metre a year. So in situation like Mill Hill, where the new woodland is right up against much older woodland, the existing populations should have been able to make slow headway and steadily colonise the newer habitat. What it has done, though, is to run at speed along the whole length of the woodland at Mill Hill: to have achieved this through their normal process of slow-and-steady growth, the plants would have to gone at full pelt, non-stop for a couple of hundred years to cover the necessary distances. Maybe this is what happened, but perhaps it is more likely that the plants had a helping hand from the clearing of the fallen trees in 1987, as the movement of machinery and vehicles broke the creeping underground stems into fragments, then dragged and dropped them through the younger woodland as the fallen trees were pulled out.

Wood anemone - considered a very slow coloniser of new woodland.

Bluebells, with no creeping underground stems, are still most abundant at Mill Hill in the ancient woodland at the top of the slope. But they have travelled some way into the adjacent, younger woodland, despite reproducing by seeds which are large and heavy and definitely not built for long-distance travel: the occasional plant can even be found more than 300m from the main colony. Wood anemone *Anemone nemorosa* has made it even farther, with a small patch turning up some 500m from the nearest plants in the original fragment of woodland – and this from a plant considered to be one of the slowest-spreading and so one of the best indicators of ancient woodland. Like dog's mercury, wood anemone can spread by underground stems, but that takes it just a couple of centimetres a year; movement by seed is faster, but one Scandinavian study found that the maximum rate of spread by seed seemed to be only about 85cm a year. The distances moved by plants at Mill Hill suggest a speed of double that rate or even faster. However, where dog's mercury at Mill Hill forms a single, continuous carpet, wood anemone is in separate blocks with no plants between its two locations: so this is not gradual spread but a single 500m leap. The most likely explanation seems to be that the movement was animal-assisted, perhaps in the mud stuck to the feet of a deer, or a badger *Meles meles*, or even on a human visitor's boots. Presumably, this kind of hitch-hiking would also explain the odd long-distance outlier from the bluebell colony.

Could it also be that the open spaces in woodlands have a role in helping plants to move about? At Mill Hill, a series of small glades is linked by a long, open ride running the length of the wood, and it is along this ride that the outlying patches of bluebells and wood anemones can be found. Of course, it could be that other plants have got just as far but are hidden out-of-sight beneath the trees, but it seems reasonable to suggest that most movement of seeds on animal and human feet is likely to happen where there is actually the room to move and the way is not barred by trees, scrub and brambles. For light-demanding plants like wild columbine, movement through the wood most definitely depends on the availability of open spaces, because it is only in these places that the seeds germinate and the plants flower: I remember that the first time I noticed wild columbine at Mill Hill was after the 1987 storm, when the extra light gave the necessary kick-start to long-buried seed.

The idea that ground disturbance and open space might make it easier for plants to move and spread through woodland is really just an informed guess on my part, based on looking in a pretty unscientific way at what plants have done and are doing at Mill Hill. But the ways in which supposedly slow-moving ancient woodland plants are able to travel through the landscape seem relevant to the ongoing debate about what the original European wildwood looked like. There are some who suggest that the woodland of Mesolithic Britain and Europe, before the arrival of farming, was an almost continuous tree canopy, with any open space being small and localised. On the other side are those who suggest that it was a dynamic mixture of open space, scrub and dense tree-cover, maintained by and changing in response to the movement of wild herbivores – wild boar various species of deer, wild cattle, wild horses. There are strong arguments on both sides, but I wonder whether the fact that so many ancient woodland plants got to Britain at all is a strong argument in favour of there having been a large proportion of open space maintained by large herds of wandering herbivores. We see ancient woodland plant species as slow moving, yet they must have really got a wiggle on as they moved up through Europe after the last Ice Age: if they had not done so, then they would never have got to Britain before it was cut off from mainland Europe by rising sea-levels. The author of the Scandinavian study into movement by wood anemone points out that this particular species can only have reached Scandinavia at all by moving at an average rate of several hundred metres a year – many, many times faster than it could have managed by seed alone – and a similar rate of movement is no doubt true for Great Britain. Alright, wood anemone has moved 500m from point-to-point at

Mill Hill, probably by hitching a ride in the mud on something's or someone's foot, but the new outlying colony is still small and its establishment may have taken years. To get the kind of speeds needed for wood anemones to have reached North-west Europe, there would need to be a lot of these kinds of little colonisation events, and that would mean lots of animals moving large distances: that kind of movement must require the necessary space for animals to move through.

What this all suggests is that even ancient woodland plants, which are generally sedentary, can move around and find their way to new places if they are given the right conditions and enough time. And time is important: the experience at Mill Hill shows that bluebells, wood anemones and other species can make big leaps into younger woodland, but it has taken 250 years for even this limited progress. There are plenty of other woodland plants at Mill Hill which are still stuck back at the starting line, confined to the original small patches of ancient woodland. Toothwort *Lathraea squamaria* is one: this strange, parasitic plant grows on the roots of trees such as field maple, hazel *Corylus avellana* and wild cherry. At Ranscombe Farm Reserve, it is a springtime regular at the very top of Mill Hill, poking up its spooky, fleshy, pink spikes of flowers often as early as March – when some early emerging bumblebees are grateful for the nectar – but it has yet to move into the younger woodland. Neither has goldilocks buttercup *Ranunculus auricomus*, very much a plant of ancient woodland. Both these species seem likely to be producing seed every year, yet they stubbornly refuse to shift themselves.

The truth is that though long-distance movement by any ancient woodland plants must in theory be possible, it is still dependent on lots of factors coming together. There must be lots of suitable habitat in the landscape, so that there is a high chance that seed will land in a good spot. Plants must be able to hitch a lift on an animal that travels long distances through the landscape, as wild cattle, horses, deer and pigs would have done in prehistory. And, finally, given the very small chance of success – of getting a lift, landing not just in the right habitat but in a place where the seed can germinate, and the resulting plant living long enough to produce seed and create a new colony – the number of seeds available to start with must be huge. The smaller the chance of any of these things happening – seed being produced, or transported, or landing in a good spot – then the longer it is likely to take for any movement to occur.

Today, not only are Britain's remaining ancient woodlands small and isolated, but the big roaming herds of large wild animals have long gone, as have (mostly) the traditional farming processes which may have replaced them, such as the grazing of extensive areas of unfenced common land or the droving of livestock (i.e. moving animals long distances on foot). The

The parasitic plant, toothwort, puts up spikes of pale pink flowers in early spring. Here is is growing at the top of Mill Hill. Photograph: R I Moyse/Plantlife.

landscape is no longer present and the ecological processes no longer operating that once allowed ancient woodland species to find and occupy new areas of woodland. This is why ancient woodland plants are just that: plants only found in ancient woodland. They arrived in those places long ago, when the systems were still in place which allowed them to move through the landscape.

This is the reason why ancient woodland can only be defined as woodland present since at least 1600AD, and not, as sometimes happens, woodland over 400 years-old. Ancient woodland is only woodland over 400 years-old because currently it is about 400 years since the year 1600. Without the kind of landscape and ecological systems that allow plants to find their way into newly created woodland, our ancient woodlands will still look distinct from other woodland in another 400 years' time. As Mill Hill Shows, even given 250 years, some ancient woodland plants do not move into younger woodland when it is right next door. Nobody really knows how long such movement would take, and what processes would be needed to make it possible. Whatever some politicians might like to think, ancient woodland is something that just isn't being made any more.

Today, ancient woodland occupies only around 2.5% of Britain's land area. We don't know whether we can recreate it or how long it would take, so what's left should surely be considered sacrosanct. Yet ancient woodland is still torn down and dug up for building and infrastructure projects. When the Channel Tunnel Rail Link (now known as HS1) was constructed in the 1990s, it tore a swathe through the ancient woodland of Cobham Woods, despite the woodland already having been designated a Site of Special Scientific Interest. As part of the mitigation of this impact, there was new woodland planted in what is now Ranscombe Farm Reserve, linking up blocks of ancient woodland, and using translocated soils from woodlands that were lost. After more than twenty years, the newly planted woodland still looks like newly planted woodland, though bluebells have made it at least 25m down into the new woodland, and a few other ancient woodland plants, like dog's mercury and barren strawberry *Potentilla sterilis* have found their way in, albeit in small amounts. This progress is certainly better than nothing, and hopefully the planted areas will eventually – perhaps in a couple of hundred years – look similar to the ancient woodland they are intended to replace. But this still ignores the depth of history and the unfathomable complexity of ancient woodland. If they tore down Rochester Cathedral and replaced it with a concrete facsimile, then that replacement might be better than nothing at all, and given centuries of time and use, it might accrue at least some of the same layers of meaning of the original. But how could such a replacement ever really make up for what has been lost? And what happens in the meantime?

Ancient woodlands like those at Ranscombe Farm Reserve provide an interesting tension. The best ones have been around as woodland for many centuries and stand like unchanging ancient monuments – indeed, in many senses they are ancient monuments. But the species which they contain all came together through a set of dynamic processes that carried them hundreds of miles northwards as the great ice sheets retreated, and the future of woodland in Britain may depend on restoring at least some of that lost dynamism. The trees we love best are the great, old giants that have stood for hundreds of years, yet the flowers we enjoy most are bluebells and the others which fare best where woodland is coppiced – and coppicing is perhaps a process of 'unwilding' a woodland by treating it fairly intensively as an economic resource. Though, yes, there is an historical precedent for this which goes as far back as the Neolithic, when people found that most trees, when cut down, quickly resprout from the stump, producing several new stems, and that this process can be carried on almost indefinitely: a coppiced tree can live far longer than it would if left uncut.

It is also absolutely true that the regular input of light into the wood, when the coppice is cut, boosts the growth of woodland plants like bluebells and early-purple orchids, which then flower more strongly in the following years, and the dense young regrowth is good cover for many bird species and is famously a good habitat for the hazel dormouse *Muscardinus avellanarius*. But because coppicing became, over the centuries, the predominant form of woodland management in Southern Britain, this meant that most woodland trees were very young – or at least their stems were, even if their roots were much, much older. In many cases, a number of larger, single-stemmed trees, or standards, were left, but these were again for economic use as larger construction timbers and wouldn't live out their full life-spans. Most of the really large, old trees that remain in the British landscape are found outside woodland: for example on grazed commons or landscaped parkland managed as wood-pasture where trees were dotted across open, grazed land. Inside woodlands, the oldest trees might be just those few left deliberately as obvious boundary markers.

When the economic value of coppicing collapsed in the mid-20th Century, it left huge areas of woodland in a kind of limbo, without the big ancient trees of a wild wood or the extensive open space of a managed woodland. Small wonder, then, that much woodland conservation management has focused on restoring active coppicing, especially where woods had been replanted with sweet chestnut which grows so very rapidly and casts such a dense shade. But this can draw attention away from the value of trees at the other end of the age scale: the big, ancient – even dying – old trees that, through their great age, act effectively as a whole ecosystem, with the branches supporting slow-growing lichens, fungi living in and on the abundant dead wood, beetles tunnelling inside, and birds nesting in hollows and cracks.

At Ranscombe Farm Reserve, most of the woodland, which is very much dominated by sweet chestnut, is actively coppiced, though with deliberate efforts to slow the regrowth of chestnut in favour of a mix of other self-sown trees and shrubs, including oaks, hazel, field maple and hornbeam, in order to create a more diverse and natural woodland. Those large, old trees that remain are retained, and where possible have been 'haloed' by clearing some of the taller coppice trees immediately around them: this creates the less shady but still sheltered conditions which can help prolong the trees' lives into great old age. In places, Cobham Woods still has some massive old trees which once would have acted as boundary markers between one person's land and another, or, in the case of one huge sweet chestnut, probably the boundary between the three local parishes of Cobham, Cuxton and Strood. All such old trees are retained unless they are a real risk to visitors, and their replacements, in the form of young trees, guarded and cared for.

Retaining and caring for old trees, maintaining the process of coppice management, allowing a wider range of tree species to establish, and creating and managing rides and glades are ways of addressing the difficult tension between the dynamic and the stable aspects of woodland. Perhaps there has been too much emphasis on coppicing as a way of managing woodlands for wildlife, but the driver behind it was a reasonable one: conservationists in the second half of the 20th Century saw how coppicing was declining as a form of woodland management, and how this was leading to declines in associated wildlife. We may want, as an alternative, to restore woodlands to a state similar to the wild wood of Mesolithic Britain, but we can only get glimpses of this distant past and cannot fully understand how these long-lost woodlands worked and what we may have lost – or gained – since. All we can do is try to manage the woodland in a way which works as well as possible for the plants and animals which we value. Some of these want their woodland young and open, others want it old and shady and stable. Some almost certainly want processes of change and movement that no longer exist. Probably every aspect and every stage of woodland has its own associated wild species, and some of the best illustrators of this may be orchids. Which brings me to the next chapter.

Chapter 6 - Tramps and thieves

Somewhere in the world of orchid studies, there must be room for a paper on why orchids fascinate people so much. Not that I'm suggesting that people shouldn't be fascinated by orchids, because I am not unaffected by them myself. I still remember my schoolboy excitement of finding a common spotted-orchid *Dactylorhiza fuchsii* on the side of the A228 in Cuxton – more or less opposite Ranscombe Farm, in fact – on a summer's day when I'd decided to walk home from school in Rochester. It was an exciting thing to find for a budding teenage botanist, and, more than 40 years and many hundreds of common spotted-orchids later, they are still pleasing to look at. But why this interest? Why does a jaded old soul like me still go, 'Ooh, an orchid!' when I spot one? Why is it that more-or-less everyone's ears prick up when you start talking about them?

There is, I suppose, their perceived rarity, and some are certainly very rare, though others are widespread and relatively common: my common spotted-orchid for one. And look at pyramidal orchids *Anacamptis pyramidalis*, which have undergone a recent expansion and spread which means they are all over the road verges and roundabouts of Southern England. Perhaps more than rarity-value is the sense that an orchid is something exotic, possibly because we are more familiar with the tropical species and hybrids in garden centres than we are with our home-grown ones. Indeed, when I googled for a picture of an orchid, I found I had to scroll past more than a hundred tropical (though undoubtedly photogenic) orchids before hitting a British species. Though, having done just that, the first British orchid in the images that came up for me was a bee orchid *Ophrys apifera*, and these do have a hint of the exotic about them, with that fat, fuzzy, brown lower lip beneath pink and spreading sepals.

Common spotted-orchid.
Photograph: R I Moyse/Plantlife.

Bee orchids are certainly not alone amongst our orchids in having unexpectedly colourful or extravagant flowers, but equally there are British orchids which are quite plain, small-flowered and dull-coloured and have none of the visual excitement that many people would expect (though because this can make them harder to find, it does give them a different sort of appeal for the dedicated plant hunter). Whatever the attractions of British wild orchids, I've always felt, whenever I've spoken to people who hadn't realised that we had wild orchids in the UK, that I need to lower their expectations in case they expect to see something like one of the big hot-house varieties with their sprays of huge, colourful flowers.

But if the appeal of orchids is less about rarity and beauty and more about their diversity, complexity and sheer weirdness, then that is more obviously understandable. They are definitely odd plants. This isn't the book for going into the detailed biology of orchids – there are others to read if you want to explore that – but what do you make of plants which cannot germinate unless the seed forms an intimate relationship with a soil-living fungus? Where it can be literally years before a seedling produces its first leaf? Where some plants co-opt fungi as accomplices in stealing nutrients from nearby trees? Or where each individual flower turns upside-down before it opens? And that's just for starters. There is also the weird variety of ways in which orchids persuade insects to pollinate them, which might include offering a decent reward of nectar, or alternatively just conning insects by just pretending to have nectar or even being a big old flirt.

It has taken orchids a good, long time to get this weird. Flowering plants first evolved more than 130 million years ago, with the earliest orchids appearing some 100 million years back (which, incidentally, means that the evolution of orchids and the formation of the geology that underpins Ranscombe Farm Reserve have been going on for about the same length of time: flowering plants really took off during the Cretaceous period, the same period when chalk was being laid down). They've now changed so much from the original flowering-plant blueprint that they are regarded as being amongst the most highly-evolved plant families. Though remember that being highly-evolved is not always the same as being more complex, and certainly does not mean better. Evolution is not directional in this sense, and evolutionary success for a species is no more than an increased ability to reproduce and increase the abundance of genes it carries. This might mean becoming more simple in design or behaviour: while orchid flowers have become highly complex and specialised, grasses, which are considered just as highly evolved as orchids, if not more so, have gone the other way: their flowers have become simplified and reduced. Or evolutionary success could simply mean staying more-or-less the same. Buttercups are much, much closer than orchids to the original blueprint for a flowering plant, but are still going strong: they have been evolving for as long as orchids, and have faced the same ups-and-downs of the last 100 million years or so. It just happened that buttercups found a good design early on and that design has been good enough to last.

Nonetheless, the weird complexities of orchid lifestyles have made them hugely successful: the orchids are probably the most diverse plant family on the planet, with more than 25,000 different species worldwide, and they are found on every continent except Antarctica. Out of this huge total there are 56 species of orchids which are found in Britain. Or 51, depending on what book you look at and how you define a 'species', which turns out not to be as easy as you might think: in common with some other plant groups, orchids don't seem to like being put into the neat boxes we like to make for them. Fortunately for us, the eleven species of orchid found at Ranscombe are not amongst those that botanists like to sub-divide and shuffle around every so often. We know what we're dealing with here, and they are a mix of the common and the somewhat rare, the showy and the not-so-showy, the nippy and the slow to move. But all have that orchid inscrutability: the life of an orchid is so complicated and so much of it is hidden that understanding them enough to conserve them is a real challenge.

For the orchid enthusiast, there's a definite orchid season in Britain, with the different species appearing and flowering in a series starting in April and going through to August. It is somewhat like butterflies, where each succeeding month brings something new, and perhaps that also helps explain people's enthusiasm for orchids: there's always something new to look forward to, right through the spring and summer. It also helps me as it gives me a framework for talking about the orchids at Ranscombe, which start with early-purple orchid in the early spring and end with the final flowering of broad-leaved helleborines *Epipactis helleborine* in August.

Early-purple orchids are up and out with the bluebells in April, though they first become apparent some weeks earlier as rosettes of purple-blotched leaves. This is mainly a woodland species in Kent, growing as much as a foot or so tall and producing fine heads of purple – or very occasionally white – flowers. It is a plant that remains common and widespread in Britain, being found from the tip of Cornwall to the top of Shetland (where the wind is punishing and the plants tiny), and it is scattered across Ranscombe's woodlands on the clay soils. The numbers of visible plants varies from year to year, as they are one of the woodland plants which get a boost from coppicing: clearing the trees allows in more light and gives them a temporary local boost, after which the numbers of visible plants decline as the trees grow back and the shadier conditions are restored. Nonetheless, there are always plants to find every year, especially in Merrals Shaw in the south-east corner of the reserve, or close to the footpath along the southern edge of Cobham Woods. Indeed, it is in the

Early-purple orchids - the first orchids of the year to flower at Ranscombe Photograph: R I Moyse/Plantlife.

Ranscombe area that the plant was first reported for Kent, in a book called *Botanologia*, published in 1664 by one Robert Turner. He apparently wrote that enough early-purple orchids grew in Cobham Park to "pleasure all the seamen's wives in Rochester" – though this does not mean that these women necessarily loved flowers but relates to the use of orchid tubers as the basis for an aphrodisiac (but please don't try this at home). This link between orchids and sex is not a completely random association but comes from the shape of orchid tubers: in many orchid species, this underground part of the plant consists of a pair of roughly egg-shaped bodies; in fact, the word *orchis* is Greek for testicle. It may also say something about Robert Turner's own preoccupations, though plants and sex seem to go back a long way and to have been on the mind of a lot of early botanists, both where the connection is fairly obvious (the cylindrical central part of the flower of lords-and-ladies or cuckoo-pint *Arum maculatum*) or obscure (in The Englishman's Flora, Geoffrey Grigson finds ancient sexual connections with the demure-looking cuckooflower or lady's-smock).

Next to flower, in May, is a series of orchids which are rather less common and rather more threatened. These are white helleborine *Cephalanthera damasonium*, lady orchid and fly orchid, all of which are on the British Red List as 'Vulnerable', and man orchid, which is considered to be at even greater risk of extinction and therefore classified as 'Endangered'. Man orchid is a cracker of a plant. Not the biggest and showiest of the orchids, quite often they are only about 15 cm (six inches) tall, sometimes more – maybe up to 20-25 cm (eight or ten inches) – or sometimes even smaller. But the flowerheads are well supplied with a good number of smallish, yellow flowers, often tinged with red, each of which really does look like a child's drawing of a little stick person, complete with arms and legs and an outsized head. In Britain, most man orchids are found on the North Downs of Kent and Sussex, where they are very much plants of chalk grassland, though often doing well where there is a bit of scrub.

The first time I saw man orchid at Ranscombe Farm would have been back in the late 1970s, when I found a few up in the woods at Mill Hill (they can cope with some shade, though they prefer the sun), though I am pretty sure they weren't in the grassland at Mill Hill at this time. By 2010, when Plantlife started counting them, there were three discrete populations: seven plants flowering in the grassland at Mill Hill, as well as 85 at the western end of the reserve in the grassland above Kitchen Field, and four on the bank by the car park at the reserve entrance on the A228 Sundridge Hill. The car park bank is interesting, because man orchids do turn out to have a bit of a thing for growing on road verges more generally: across the North Downs, there are quite a few steep, chalky roadside banks which have their own populations of man orchids.

But the plants on the bank at the Ranscombe Farm Reserve car park illustrate the vulnerability of such populations to climate change. In recent years, there seems to have been an increased frequency of warm, dry weather in the early spring, a time when man orchids already have their leaves above ground and are starting to produce flower-buds. When this weather is particularly prolonged, we have seen emerging plants abandon the effort of flowering, shrivel up, and retreat back into their underground tubers. This has even happened on the deeper soils of the grassland at Kitchen Field, though here those plants growing in the partial shade of the trees are more easily able to continue flowering, protected from the worst effects of the strong spring sunshine. Most of Britain's man orchids are found on the south-facing, open chalk grassland of the North Downs, which for centuries provided the warm, sunny habitat which favoured a plant at the northern edge of its European range. But if climate change means that dry early spring weather becomes a regular pattern, then their previous home may become increasingly inhospitable.

The retreating back into the tubers is a handy strategy, though. Quite a few orchids seem to do this, and for man orchids it is part of their way of life. One study found that the proportion of man orchid plants flowering in any given year varied widely over a thirteen or fourteen year period, from less than 20% to more than 70% of the local population. The man orchids at Ranscombe Farm certainly seem to disappear from view on a fairly regular basis, especially after a particularly good flowering year: it looks as if the effort of producing flowers and seeds is so great that the plants need to rest for a year or two so that they can recuperate and build their strength up again. Over a twelve-year period from 2010 to 2021, even as the total man orchid population at Ranscombe quadrupled from just under a hundred flowering plants to more than 380, there was a big dip in numbers every three years. So far, the pattern has repeated three times: the number of flowering plants increases for two years, and then drops, in one case by nearly three-quarters. This tells you two things: it's tough being a man orchid and flowering takes effort and resources; and it takes years of monitoring a population to see whether it's really growing or shrinking or just having a bad year or two. It also means you cannot base any decisions about habitat management on small amounts of data. Some early tree clearance work was followed by an immediate and substantial drop in the size of the Kitchen Field colony of man orchids – with numbers taking three years to recover – but ten years later this particular population had nearly tripled in size.

Overall, at Ranscombe Farm Reserve there is enough information to show a pretty steep and continued growth in man orchid numbers over the last seven or eight years. This has been the result of targeted work, though

Man orchids quickly colonised the restored area of grassland at Mill Hill. Photograph: R I Moyse/Plantlife.

it is hard to pinpoint exactly which individual actions had what impacts, again because of the strange orchid lifestyle. All orchids are unable to grow without the help of soil-living fungi. The fungi infect the germinating seed and the young seedling as it develops into an underground structure called a 'protocorm'. Or perhaps it is more accurate to say that the orchid infects the fungus, as the young seedling lives off the fungus for more than a year, stealing from the fungus to fuel its own growth before putting up its first leaf at least a year-and-a-half after initial germination. Even after this time, when the orchid is able to produce some its own food through photosynthesis, it is thought that it gives nothing back to the fungus. It is then even longer – a couple of years or perhaps more – before the plant produces its first flowers. This long delay means that where you get a visible increase in the numbers of plants – which usually means spotting the flowers – this will most likely be the result of something that happened or some bit of habitat management work as long as three or four years before.

You can draw some conclusions, though. For instance, clearance of scrub that was shading the grassland at Kitchen Field led to an immediate increase in flowering by the existing man orchids – and in the size of the flowering plants – clearly showing that the additional sunlight had boosted the growth of plants already present, stimulating them to make more flowers and so more seed. And removing dense, 60-year-old scrub to extend the tiny patch of grassland at Mill Hill allowed plants to spread into the newly restored grassland – and to flower just four years after the new habitat became available. Overall, the management that seems to be working best for man orchids at Ranscombe Farm Reserve is cutting, followed by grazing with cattle, in autumn, and this has produced a steady increase in the number of plants above Kitchen Field. The use of cattle to graze the grassland may – just may – have also been instrumental in getting man orchid into the large area of restored grassland in Brockles Field.

Brockles Field is an area of grassland of 23 ha (some 60 acres) which was bought by Plantlife in 2010 as an addition to the reserve. It is managed mainly by grazing with cattle, and, in some years, cattle have been moved directly into Brockles from Kitchen Field, following the autumn grazing there: because the fields are next to each other, it is easy to get a few volunteers together and walk the animals through. In 2020, man orchids were discovered in Brockles Field for the first time when a patch of 39 plants was found flowering some 600m from the nearest population in Kitchen Field, and they have clearly taken to their new home, as three times this number was counted in 2021. A bit of guesswork about timings, and a look at the records of cattle movements, suggests that the origin of this population might well have been cattle carrying seeds in the mud on their feet as they walked between the two fields. Sadly, though, this is impossible to prove one way or the other, as orchid seeds are minute and easily small enough to be carried on the wind,

Though just because a plant has tiny, light-weight seed doesn't mean that seed will necessarily be blown far and wide. This is something that is shown very clearly by lady orchids. Now, I said a lot earlier about British orchids by-and-large being not especially large and showy, but the lady orchid is one of the exceptions and is a plant that definitely will not disappoint anyone coming fresh to British orchids. It is big and bold, sometimes two feet tall, with broad, glossy leaves and a large flowerhead which is white, pink and dark red like a good serving of ice-cream with raspberry sauce on top. The name comes from the individual flowers, which are meant to resemble little ladies in crinoline dresses and pink bonnets, though this takes a bit of imagination, frankly. However, the overall effect of these plants is assertiveness: here I am, they say, and what are you going to do about it? (To which the answer must be, "I'm not going to do anything. Please give me more.")

There are only small numbers of lady orchids at Ranscombe, but they are worth having. These are very much Kent plants, being almost entirely limited to this county's part of the North Downs (though with a few outlying populations as far as Oxfordshire), where they are mainly found in open woodland, wood edges and scrub on the chalkier soils. At Ranscombe, they are found in two locations: in the south-east of Cobham Woods, growing in ancient coppice woodland around the top of a shallow valley where chalk is relatively close to the surface; and on the distinctly chalky soils at Mill Hill, where a few plants are found under the young, scrubby woodland inside the reserve boundary and a larger population in similar habitat just outside.

A lot of work has gone on at Ranscombe Farm Reserve to try and persuade lady orchids to increase in number. Fortunately, and unusually for almost any plant, there's lots of available scientific information on them, largely thanks to Hans Jacquemyn and others at the Catholic University of Leuven in Belgium. It is from the work of these researchers that we know that, even though lady orchid seed is as fine and light as any other orchid seed, it doesn't seem to ever be carried far on the wind, and most of it seems to fall within four or five metres of the parent plant. This is not just down to lady orchids living amongst sheltering trees and scrub: one study of orchids of open habitats found that even amongst low vegetation, the proportion of seed falling beyond about 2.5m (around 8 feet) was so small as to be unmeasurable. It seems to be a fact of physics and probability that, although very long-distance travel by orchid seeds can and does happen, it is a very rare event. And this doesn't just apply to orchids, but to all sorts of less mobile species. Some of the more sedentary butterflies show the same effect: if you plot a graph with distance travelled as the horizontal axis and the number of individuals which travel that distance as the vertical, then it is very strongly weighted to the left, with most individuals hardly travelling anywhere, but with a long, low tail to the right which indicates a very few individuals covering much, much longer distances. From the point of view of trying to conserve such species and getting them to move naturally and colonise new sites – especially important in light of climate change – the issue of probability is all-important. If there is only a tiny chance of any one individual seed moving a significant distance from its parent, then you need to have a big population of parent plants, all making lots of seed, to have any kind of realistic chance of this sort of movement happening at all.

Of course, you would think that this wouldn't be a problem with a good-sized orchid, which might have dozens of flowers each capable of producing many thousands of seeds, but this turns out not to be the case. The chance of an orchid seed landing on a suitable spot to germinate, finding the right soil-living fungus, and then making it to adulthood is absolutely tiny. So much so, in fact, that some of the Belgian studies found

Lady orchids in Clay Pond Wood, growing around an active badger sett.

that the continued survival of lady orchid populations is dependent on the enough flowers being produced and a high enough proportion of these getting pollinated and going on to produce seeds. Producing tens of thousands of seeds may simply not be enough: it may be necessary for a population to regularly produce hundreds of thousands of seeds just to stand still in terms of overall numbers. Getting a lady orchid population to increase means persuading as many of the plants to flower as possible, and then trying to ensure that as many flowers as possible are pollinated. Getting the plants to flower is relatively easy: plenty of light seems to be key – producing a flowerhead is an effort for the plant, so get lots of light to the leaves and the underground tuber can put on growth and store up food reserves. But pollination is different matter, because lady orchids turn out to be in the business of conning insects into doing the job for nothing. The flowers look and smell as if they contain plenty of nectar, but this is a lie and they are nectar-free – and any visiting insect quickly finds this out, though not before it has picked up the load of pollen. Then the insect has to fall for the same trick a second time in order to pass that pollen on to another flower. As a strategy, this clearly has a downside for the orchid.

because it turns out that in practice, natural pollination rates are generally low in lady orchids, with studies reporting maybe ten or twenty per cent of flowers generally getting pollinated. At Ranscombe Farm in 2021, a count of more than 600 individual flowers on thirteen different plants in one particular patch found the proportion of flowers pollinated on each plant ranged from none at all up to around 28%, with an average of less than 7%. One plant was found where just under 30% of the flowers on the plant had been pollinated: but this was a plant in a separate population located in an open bit of woodland where there were plenty of other flowering plant species present, suggesting that perhaps the more generally flowery environment pulled in a lot more insects, and so, perhaps, more unsuspecting customers for the lady orchid to work its con trick on. But, even so, these pollination rates are much lower than in those orchid species that are prepared to offer some real nectar as a proper reward.

A lady orchid on its own, as in the case of one large plant that grew in isolation for seven or eight years in a section of woodland at Mill Hill, seems to have little chance. It is the case that lady orchids are self-fertile, so pollen from one flower on a lone plant could potentially pollinate another flower on the same plant, and some seed be produced (though self-pollination does apparently make for lower quality seed with less chance of successful germination). But although the plant flowered every year for at least nine years, no self-pollination occurred and it produced no seed at all until it was deliberately hand-pollinated using pollen from a nearby population. The fact that no natural self-pollination occurred may be due to the weird way in which lady orchids and many other orchid species package up their pollen for delivery. Rather than scattering the pollen freely in the way lots of other plants do, these orchids have special bodies called pollinia which each consist of a little ball of pollen grains attached to a short stalk; at the other end of this stalk is a sticky pad. The pad attaches to the head of any insect which comes visiting in its search for nectar, and off goes the insect wearing these two club-shaped objects like a bit of novelty headgear. In order to pollinate another flower, however, the pollen needs to be in a slightly different position so that it can hit the right receptive surface inside the flower itself. So each of the pollinia bends forward on its stalk, a process that takes a few minutes – by which time the unlucky insect may have realised it's not going to find any nectar on this particular plant and moved off.

Hand-pollination of the lonely lady orchid meant using cocktail sticks to stand in for the front end of a visiting insect: poke one into a flower in the right way, and it comes out with pollinia attached. After a short while, the pollinia bend forward and it is fairly easy to push the stick into another flower so that pollen is deposited in the appropriate spot. So, sticks in hand, we went about the job of pollinating our lonely lady orchid, and it worked very well, producing nice, swollen seed pods on every flower we worked on. This has been repeated a couple more times since, aided by the recent appearance nearby of a second plant which made hand-pollination easier, as pollen-loaded cocktail sticks no longer have to be carried more than 500m through the woods from the nearest other plants. The big result is that the two flowering plants have very recently been joined by a little group of seedlings. Which means that barring accidents (or plants being eaten by rabbits *Oryctolagus cuniculus* or deer or slugs), the tiny population at Mill Hill has started to grow. Young plants and seedlings have also been spotted in the other populations within Ranscombe Farm Reserve, but measurable growth is always going to be slow, with lady orchid plants taking three or four years before producing a single leaf, and maybe eight or ten years from germination to their first flowering.

Lady orchids aren't the only orchids to cheat their pollinators. Fly orchids go one-step further by offering not fake nectar but fake sex. At Ranscombe Farm Reserve, fly orchids can be found under scrub and young woodland in a couple of locations, where they push up spikes of small, brown flowers that can be hard to spot in the shade. The lip of the fly orchid flower is drawn out into an elongated shape which roughly resembles a smallish insect, even to the extent of having a couple of short 'antennae' at the top, hence the plant's name, though the insect which it is impersonating is a kind of wasp rather than a fly. The likeness is pretty crude to human eyes but, coupled with the appropriate chemical scents, it is enough to attract males of a species of digger wasp. Sometimes called the field digger-wasp, two-girdled

A male digger wasp, *Argogorytes mystaceus*, is fooled into trying to mate with a fly orchid flower. Photograph: Sue Skipp.

Fly orchids can be really difficult to find in the woodland shade.
Photograph: R I Moyse/Plantlife.

digger-wasp or cuckoo-spit wasp, *Argogorytes mystaceus* is one of the many species of solitary wasp which do not have a worker caste and where individual females work alone to excavate the burrows where they lay their eggs. Each egg is provided with a store of prey insects, which the wasp larva will feed on to fuel its growth to adulthood. In the case of *Argogorytes mystaceus*, the prey are the froghopper nymphs which create and hide inside the frothy blobs on plants which we call cuckoo-spit. The male wasps play no part in digging burrows or rearing young, but spend their time looking for females to mate with, and they cannot distinguish between the sight and smell of a fly orchid flower and those of a real female wasp and the pheromones she produces. In trying to mate with the flower, the wasp picks up the pollen, which is transferred to another flower the next time he makes his big mistake.

Trying to ensure that fly orchids increase and spread at Ranscombe therefore means encouraging plants to flower and making sure that conditions are right for the wasp, as no other insect can do the job. Our plants were initially in a good deal of shade, and digger-wasps need warmth and sunshine just to stay active, as well as sunny spots where they can dig their burrows. So thinning the tree cover was the first thing to try, and we were fortunate to have the help of an undergraduate student, Stephanie Skipp, who was working on a dissertation on how light availability affected pollination and fruit-set in fly orchids. What was found was that the increase in light getting to Ranscombe Farm's fly orchids led to increased flowering, with plants making more flowers on taller stems, but that it didn't make any difference in the rate of pollination by the wasps. There are several possible reasons for this, but my suspicion is that it is because Ranscombe Farm's fly orchids are in parts of the reserve where there was already a good mix of shady woodland and sunny open space. For instance, at Mill Hill, the wide woodland ride and series of glades must provide lots of opportunities for male wasps to warm up before heading into the shade for their orchid liaisons, so there is no need for the plants themselves to be in full sunshine. By contrast, when Steph Skipp looked at other Kent sites where there was less open space overall, she found that the light level around the fly orchid plants did make a difference to pollination – so pollination of fly orchids is perhaps less dependent on the plants themselves being in the light, so long as there are plenty of well-lit, open spaces in the general area.

We also found an unfortunate and unexpected side-effect of thinning tree and scrub cover – the fly orchids may have flowered much more, but they also suddenly became very visible to rabbits, which started eating the flowerheads off plants which they had previously ignored. This seemed to lead to the rabbits getting a bit of a taste for the orchids, and the following year they ate just about every fly orchid plant they could find. Fortunately, this was remedied fairly cheaply and easily by putting up some basic rabbit-fencing to keep the animals off the area with the greatest concentration of fly orchids – and, strangely, this seemed to keep them off those plants outside the fence as well. Perhaps the local rabbit population had gone through a sudden crash; or perhaps seeking out and eating fly orchids was a habit they developed and then just as quickly dropped when their main supply was cut off.

Another orchid which rabbits also sometimes seem to get a taste for is the white helleborine, which flowers around the same time as the fly, lady and man orchids. White helleborines are often seen with their flowerheads nipped off, but when they do get to flower, they are supremely elegant and lovely plants. Oval leaves are arranged alternately up a single vertical stem which can be up to a foot or more tall. At the top are up to a dozen (though often just a few) well-spaced flowers. The flowers are a creamy, candle-wax white and generally open only slightly, so that the slender, white flowers and the neatness of the whole plant puts me in mind of a candelabrum. The resemblance is apt, because white helleborines do well where there is little light. They are plants of shade, and often deep shade. At Ranscombe Farm, they can be found under the beech trees at Mill Hill and under hornbeams and oaks above Kitchen Field, and their sneaky technique for thriving where other plants would be shaded out is simple: they steal.

Like all other orchids, white helleborines are dependent on soil-living fungi, which grow into their roots, and they take many of the nutrients they need from the fungi. But they take things a step further than most other

White helleborines growing beneath - and making a living from - a mature beech tree at Mill Hill. Photograph: R I Moyse/Plantlife.

orchids, as they associate with species of fungi which also have links to beech trees (and sometimes other tree species, as at Ranscombe Farm). So the relationship is a sort of *ménage-à-trois*, with the fungus in the roots of both the orchid and the tree – but it is an unequal partnership. The relationship between tree and fungus is a two-way thing: the fungus passes minerals to the tree and gets energy-rich carbon compounds – the products of photosynthesis – from the tree, so both benefit. The orchid, though, takes minerals and carbon compounds from the fungus, and seems to give nothing in return. This is how white helleborines do so well in deep shade: they are freeloaders, living on all the photosynthetic work the beech trees are doing. Though the relationship between tree and orchid could also be put another way. After all, the beech trees are using their ability to grow tall and cast dense shade in order to destroy any plants which might otherwise compete with them for space. For a plant, a leaf is not just a solar panel for capturing energy from sunlight and fuelling its own growth: it is also there to cast shade on other plants nearby and so reduce its competitors' own chances of growth and survival. With this in mind, the white helleborine's approach of making its own quiet way by stealing a little from the huge beech – though without ever challenging the tree's dominance – does look more like an act of admirable cunning and something worthy of the central character in an old folk-tale hero or even Hollywood movie.

Bird's-nest orchids live entirely off their fungal associates, and so have no need for chlorophyll.

White helleborines are not alone in this approach of using the fungus as an unwitting stooge in stealing from a nearby tree. Bird's-nest orchid *Neottia nidus-avis*, which is very rarely found at Ranscombe Farm Reserve, not only uses the same method, but has taken it to its extreme: it has no leaves or chlorophyll of its own and gets all the nutrition it needs via the fungus. It is effectively a parasite on the fungus and tree, and, without any need for sunlight, only needs to stick its head above ground when it wants to flower. This means that it can potentially remain present and unseen for years: although bird's-nest orchid was previously known to be present in from Cobham Woods, I saw it only once in a ten-year period, when three plants were found beneath a large oak by a local botanist, David Steere.

Come early June, and the white helleborine and man, lady, fly and bird's-nest orchids are finishing flowering and getting ready to retreat underground till at least next year. Common twayblade *Neottia ovata*, though, seems to enjoy quite a long flowering period, so its tall spikes of small, green, unshowy flowers may be found in May and sometimes right through June. Because it has green flowers and just a single pair of large, round leaves, common twayblade can be hard to spot, but even allowing for this they seem – at Ranscombe at least – to be a bit irregular in their habits. A few plants appear every year, but seemingly rarely in the same place twice in a row: they pop up here and there in both woodland and grassland, and in recent years I've found plants tucked in amongst sweet chestnut in Cobham Woods, both in grassland and under beech trees at Mill Hill, and

Bee orchid, in the grassland above Kitchen Field.
Photograph: R I Moyse/Plantlife.

under shady scrub above Kitchen Field. This may be to do with the plants choosing from time-to-time to spend a year or two below ground, because individuals can apparently live for more than forty years.

As June progresses, and the summer starts in earnest, we begin to see the species that like to come out of the shadows and enjoy the full sunshine. At Ranscombe Farm Reserve, common spotted-orchid seems to prefer the slightly damper woodland rides, where its purple-spotted leaves might cause confusion with early-purple orchid, though its paler flowers are very different, with their patterns of pink dots and lines. Bee orchids, in contrast, like drier soil with low or open vegetation, particularly in Brockles Field or in the grassland above Kitchen Field, though they will cope with a bit of shade and sometimes turn up under scrub.

Everybody loves a bee orchid. They are noticeably different, with the fuzzy, brown lip of the flower giving it some of the characteristics of a teddy bear. Orchid-lovers are particularly drawn to the odd colour varieties that sometimes occur, many of which have been given their own scientific names. Though you might expect some odd characteristics to get locked into bee orchid populations, given that all plants in this country are self-pollinated. Although bee orchids, like fly orchids, are designed to tempt an insect to try and mate, the particular bee which pollinates bee orchids isn't found in Britain. So bee orchids use a fail-safe system where easy and effective pollination of each flower by itself means they are able to continue to reproduce without any outside help. As a result, bee orchids may look exotic, but are widespread and fairly common in England, especially towards the South-east, and can be one of the earlier colonising species of open grassy habitats, even sometimes turning up in garden lawns.

Come July, and pyramidal orchids are at their peak. This must surely now be the commonest orchid in Kent, if not the whole of South-east England, seemingly found everywhere there is a patch of rough, chalky grassland and increasing in numbers in many places. If not, then it is definitely the commonest at Ranscombe Farm, found across the site, with many hundreds in Brockles Field and increasing numbers in other grassland patches. Elsewhere, the pyramidal orchid has become a plant of road verges and roundabouts, where its deep pink flowers make it easy to spot. Incidentally, I don't know whether I'm alone in thinking that pyramidal orchid is a bit of a clunky name: one, it's tricky to say; two, it only has a pyramid-shaped flowerhead when the flowers start to open (they open in sequence from the bottom up, so the base of the flowerhead is wide and the top at first comes to a point), and, three – to be completely pedantic – it's really a cone, not a pyramid.

July is also when broad-leaved helleborines come into flower, marking the end of Ranscombe Farm's orchid season. These are handsome and quite tall plants – half a metre or around 18 inches is unremarkable, with a head of sometimes dozens of purplish flowers above a stem bearing alternating, broad, oval leaves. The woodland at Mill Hill is the place for these at Ranscombe, where they can sometimes be found in the open, close to the path and sometimes tucked away in the shade, but never commonly, and like common twayblade, they rarely seem to be in the same place twice. Anne and Simon Harrop, in their Orchids of Britain and Ireland, point out that plants sometimes spend up to three years below ground between bouts of flowering, so perhaps that explains the apparent skittering about – though the Harrops also note that this plant is a good coloniser of new sites and can take as little as 18 months from germination to flowering, so perhaps the population at Ranscombe really is one that bounces around a bit.

So, that's the tour of Ranscombe Farm Reserve's orchids. I hope it doesn't seem too indulgent of me to go through them all a species at a time, but they are all so varied and all have a different story to tell about the extraordinary and eccentric lengths that plants go to in order to survive and reproduce (and that's without mentioning that broad-leaved helleborine nectar sometimes ferments in the flower and so gets pollinators drunk). However, I do sometimes wonder whether the weird and exotic nature of orchids, their odd lifestyles, and their peculiar relationships with fungi and with trees, can get in the way of understanding what they are really about. Because they are beautiful and strange, and because in some cases they are genuinely rare, we do regard orchids with a degree of veneration, and we can imagine that the presence of an orchid means the local habitat must be a tip-top example of its kind and needs to be kept that way. But it is too easy to forget that all natural habitats change with time and always have: open ground grasses up, then becomes scrubby, then wooded. And at any point in this process, for any of a number of reasons, the clock may be suddenly reset to an earlier stage in the succession before starting over again. It is clear that all of the orchids at Ranscombe seem to want to occupy a different position in this process of ecological change. Bee orchids, with their preference for low, open vegetation, are plants of the very early stages of succession. Bird's-nest orchids are at the other end, being species of old woodland, and needing mature trees to support them. The rest – the pyramidal orchid in amongst taller grassland, the lady orchid in scrub and open woodland, and all the others – are somewhere in between the two extremes, and so must surely have evolved to take advantage of an environment in motion and undergoing change.

It is probably no accident that the part of Ranscombe Farm Reserve with the greatest variety of orchid species is Mill Hill, where you can find man orchid, lady orchid, fly orchid, white helleborine, common twayblade, pyramidal orchid and broad-leaved helleborine. Because, although Mill Hill does include a small patch of old grassland and some areas of ancient woodland, most of its woodland is less than 200 years old and was also fairly comprehensively felled by the wind in 1987. It is an environment in pretty rapid flux: for 200 years or more, it has been progressing from open ground to closed woodland, with interruptions where the ecological clock was suddenly set back by a massive storm, or on a much smaller scale by deliberate scrub-clearance, and it is within this changing environment that the orchids have found a home. White helleborines have found and taken advantage of beeches likely to have been planted in the early 19th Century, and have happily spread to exploit these trees' younger offspring, while both lady orchids and fly orchids have arrived in and colonised woodland that was completely cleared by the 1987 storm. Even man orchids, which were and still are in the remaining patch of old chalk grassland, have quickly moved into an adjacent area where grassland is being restored after decades under scrub.

I have described above the difficulties and frustrations of encouraging populations of lady orchids and fly orchids to increase and spread. Perhaps the reason that this is such hard work is because these orchids' real home is not in woodland held in a particular, steady state, but rather in woodland that is changing. Perhaps they are present at Mill Hill precisely because the trees fell over but the succession to woodland continued afterwards. And perhaps they are not designed to have stable, persistent populations but to be nomads in a dynamic landscape where the right conditions come and go in different places.

So maybe the biggest lesson from Ranscombe Farm's orchids is that the best way to restore, recreate and retain a wide diversity of plants might not be through the creation of a series of steady-state habitats – grassland here, scrub there, woodland over there – but to create dynamic ecological systems where habitats grade into each other and are under a more-or-less continuous process of change. This is not a new thought: the idea that active ecological processes are important for conserving biodiversity is part of the rationale for rewilding. But fully embracing its implications is a tough call for any organisation or person responsible for managing a nature reserve of limited size. After all, if you have a population of a rare plant or animal on that reserve, then your job is to get that population to increase and spread as much as you can: the species has its best chance of surviving on your site and of spreading beyond the reserve's boundaries only if the population is as big as it can be. Holding ecological succession at the right stage for a rare species may be difficult, but can often be made to work. In such a case, do you risk the possible decline or loss of your rare species by trying to establish natural dynamic processes that, by their very nature, have an unpredictable outcome? Or do you stick to the tried-and-tested habitat management that works but that is expensive, time-consuming and potentially not the best all-round solution?

Chapter 7 - Building grasslands

In school, we get taught a basic idea about ecological succession: that bare land is quickly colonised by herbaceous plants, that slower-growing woody plants then move in, and eventually the once-open, bare ground turns to tall, dense woodland. It is this very simple model that would have seen prehistoric Britain as being pretty much a wall-to-wall carpet of trees. But the plants tell a different story. The species of weeds of open ground that were already here when humans first arrived, the bee orchids that favour open ground, the fly orchids of open scrub – all of these and more could not have been here if Britain had been one dark, unchanging forest. The early Mesolithic residents of what became Ranscombe Farm Reserve must surely have inhabited a world where the clock of ecological succession was regularly being reset by erosion, storms, floods, and – according to some theories about the wild past of North-west Europe – by the action of large herbivores like wild cattle and wild horses, as well as by beavers. Heavily wooded though Britain clearly was at that time, it would must have been a place of continuous change as periodic creation of open ground was followed by slow colonisations by woodland and scrub.

The spread of farming and settled communities during and after the Neolithic period led to a major reduction in the tree cover and a massive increase in open habitats. But rather than this resetting the ecological clock and acting as the trigger to another round of succession to scrub and woodland, the pendulum was stilled and the clock stopped ticking. Grassland was no longer part of an ongoing process of change, but made permanent to provide a steady supply of food for livestock. Woodlands were managed to supply timber and protected from grazing animals, and the larger wild animals were driven to extinction. Arable farming also brought something completely new into the mix, with its concentrated areas of perpetual ground disturbance on a short rotation. Now grassland and arable dominate the British landscape, and flower-rich grasslands are amongst our most valued wildlife habitats. Yet both must originally have been major shocks to the system: now there were suddenly large and lasting gaps in the woodland cover. In the arable fields, whole new communities of previously unknown cornfield weeds were taking root. And, though grasslands had existed long before farming, they were now far more extensive and unchanging, and this made opportunities for grassland plants to come together in entirely new combinations.

Where Ranscombe Farm Reserve sits on the North Downs, the predominant type of permanent grassland which has survived into the 21st Century is chalk grassland or downland, the kind of dry grassland that

Greater knapweed in the grassland at Mill Hill. Photograph: R I Moyse/ Plantlife.

develops on the steeper slopes, where the soil is thin and the underlying chalk is close to the surface. To what extent this habitat existed before the advent of the Neolithic is a subject of debate: it used to be considered that the North Downs were completely wooded prior to their clearance by Neolithic farmers, though this woodland may well have had more gaps in it than previously believed. There is also evidence from elsewhere in the country that dry chalk grasslands, similar to those found today, may have been part of the landscape shortly after the last Ice Age, even as the forests were starting to dominate Britain.

Could the chalk grasslands of today's North Downs, such as the fragments found at Ranscombe Farm Reserve today, be considered as descendants of wild pre-Neolithic grasslands? Perhaps, though probably not directly. It seems that many of the commonest types of traditional farmed grasslands are not simply, as you might imagine, natural grasslands that have been put to human economic use. In his fascinating book, Meadows, the ecologist George Peterken explains how the plants that came to make up the new farmed meadows and pastures may have originally made their homes in very different types of habitat, such as open woodlands or the tall herbaceous vegetation of riverside floodplains. The result was that farming led to the development of entirely new communities of plants that incorporated aspects of a range of other habitat types.

This seems as likely to be true of grassy chalk downland as it is of the upland hay meadows and other types of grassland which Peterken deals with. For example, greater knapweed *Centaurea scabiosa*, with its flowerheads like small purple explosions, is typical of the dry Kent Downs today but around the end of the Ice Age seems to have been part of a community of tall, herbaceous plants found on wet soils: its pollen has been found in ancient wetland deposits in the Upper Thames Valley. Other plants of today's downland seem originally to have been constituents of the tundra vegetation which covered parts of Britain during later Ice Ages. Some of these low-growing plants seem to be a lot tougher than they might appear, and may well have been present in Britain while the ice went through its most recent period of expansion, when it reached as far as Southern Britain, and might even have been here before that time. Harebell, for instance, looks dainty, with its slender stems and small, blue, bell-shaped flowers, but has a current global range which extends well up into the Arctic, and was found in Britain more than 200,000 years ago. Common rock-rose *Helianthemum nummularium*, which looks so much like a sun-worshipper, growing and producing its sunshine-yellow flowers in the hottest, driest corners, was present in Northern Britain more than 9,000 years ago, just after the retreat of the last ice sheets to cover Britain. And there's evidence that horseshoe vetch, so important today as the foodplant for those heat-loving butterflies, the chalkhill blue *Polyommatus coridon* and adonis blue *Polyommatus bellargus*, may have held on in Britain right through the last Ice Age.

Wherever the plants of chalk grassland came from, they made a very rich mix, possibly because the combination of dry, shallow soil and grazing by livestock meant that no individual species could get a competitive advantage. Downland is a stressed environment, where persistence and a tolerance of poor growing conditions are needed, as are an ability to grow close to the ground to avoid being completely chewed up by grazing animals. Large thyme *Thymus pulegioides* survives by creeping and spreading along the soil surface; and its strong thyme scent is a chemical defence which discourages insects that might eat it, but also gives downland its particular smell as you walk across it. Yellow-wort *Blackstonia perfoliata* avoids losing too much water in dry, hot summers by having a waxy coat on its leaves. Dwarf thistle *Cirsium acaule* makes a rosette of sharply spiny leaves flat to the ground, with its flowers in the centre on very short stalks: hard for animals to eat – and not always easy to spot when you sit down for a picnic. Other plants avoid long-term damage from grazing animals by keeping their growing points close to the ground and just throwing up tall flowering stems in the summer. Greater knapweed is one of these, as is clustered bellflower, which sends up its deep-blue flowers, clumped tight together on the stems, in July and August, keeping its head down for the rest of the year.

Yellow-wort growing in Brockles Field. Photograph: R I Moyse/Plantlife.

Chalk grassland, with its rich mix of wild plant species (the figure of up to forty different plant species per square metre is often quoted) eventually came to dominate the steeper slopes of the North Downs. Really, this was just a matter of practicality: in a system of mixed farming, with both arable and livestock, the steep slopes are hard to plough, and ploughing them would have led to soil erosion anyway: so keep the ploughed land to the deeper, more level soils and use the dry, sloping land for the livestock. And fortunately the grassland that developed on the thin chalk soils proved to be excellent for rearing sheep. Interestingly, though this sort of system was fairly stable for long periods, some shifts would have occurred which reflected the natural dynamism of the pre-farming landscape: sudden drops in the human population due to disease, or changes in the farming economy, could lead to previously managed land being abandoned. Dan Tuson, who has studied and written about the North Downs, has pointed out that the open grassland of the Kent Downs was by no means unchanging: some areas might have cycled between arable, grassland and scrub over decades as the demands of the local farming population – and the wider economy – dictated. There would have been a similar dynamism in woodlands – not only would any woodland in active coppice management be in a constant loop from open ground to deep, shady woodland, but demands for timber would have varied with markets and with the needs of the local population.

In the last fifty to a hundred years, however, this has changed substantially. Land which can be farmed most economically for grass or arable tends to stay that way, while the less economic land – which includes coppice woodland and the less productive grassland – has been largely abandoned. In the absence of the large wild herbivores which once might have acted to reset the ecological clock, the change has been largely one way, with woodland becoming dark and dense and losing its open spaces, and grassland rapidly covered by scrub and then trees. Because of this, the process of ecological succession has been an important driver in the loss of flower-rich grassland. It is estimated that 97% or more of our flower-rich grassland has been lost over the last century or so; partly this is the result of improvement for agriculture by ploughing and resowing with more productive grasses and clovers, partly it has happened through abandonment where such improvement was not practical. Good grassland – what we might call wildflower meadow – is one of the rarest and most threatened habitats in Britain. It is certainly a lot rarer than woodland, which has substantially increased in area over the last century, even as ancient woodland has declined. Think, for example, of all those trains delayed by leaves on the line, then watch an old bit of film of steam trains: railway cuttings and embankments used to be grassy, not least because sparks from steam engines would regularly set fire to line-side vegetation. Those long, connected grassy corridors are now heavily wooded.

This is what happened to much of the chalk grassland on the North Downs when farming intensified after the Second World War: you could fatten sheep more quickly and efficiently on fertilised and resown grasslands on the more level and deeper soils than you could on the parched chalk downland slopes. This change in the economics of livestock farming meant some downland was put to arable – modern tractors and fertilisers making this now feasible – and lots just abandoned. The National Trust estimates that 80% of Britain's chalk grassland has been lost since the Second World War, and it has become a particularly threatened habitat. This is important not just because chalk grassland is so rich in wild plants (and also in the butterflies and other insects these plants support) but because it is a rare habitat at a global scale. There are few parts of the world where chalk is the surface rock, and the UK is said to hold half of the world's old chalk grassland. Around 2.5% of all the chalk grassland in the world is in the county of Kent. A tiny part of this old chalk grassland remains at Ranscombe Farm Reserve, and a lot of work has been put into maintaining and enhancing the remaining patches as well as trying to create and restore a lot more new grassland.

‘Wildflower meadow’ is a term that many people will be familiar with, and you might think it doesn’t need defining any further. But it probably is worth picking at it a bit, as there are lots of different words that are used to describe grassy or flowery vegetation – for instance, I’ve already used the words ‘meadow’, ‘pasture’ and ‘grassland’ without explaining what I mean by them. ‘Grassland’ ought to be fairly self-explanatory as any area of land where the vegetation is low and non-woody and dominated by one or more types of grass. Though, of course, wheat and barley are grasses and it is hard to see a wheat field as grassland, which means we need to qualify the description so that ‘grassland’ is defined as land dominated by one or more types of perennial grass. A ’meadow’ in the strict sense is grassland that is managed by cutting, originally to provide hay which could be stored and fed to livestock over the winter: the word itself is derived from an old Germanic word which also gives us the verb ‘mow’.

If you don’t mow a piece of grassland, then the other main way of maintaining it is as a pasture, which is grassland managed by grazing and without regular mowing (the root of the word ‘pasture’ is the Latin word for ‘to graze’). Like mowing, repeated grazing stops trees and shrubs from getting established, and encourages grasses to dominate. But it does it in a different way to mowing, particularly as animals are fussy and will choose to eat some plants but leave others, while a mowing machine doesn’t discriminate and treats everything in the same way. Even though meadows may also be grazed for a period after the hay is cut, the difference between the two types of management is such that you can often tell whether a particular grassland is managed as a meadow or as a pasture just from seeing what plants are present – and this is certainly the case with chalk downland.

Pastures and meadows don’t have to be good for wildflowers. More often than not, commercially farmed grasslands are altered to improve their productivity, for instance by spreading fertilisers, by using herbicides to favour grasses, or by sowing with a limited mixture of plants, such as perennial rye-grass *Lolium perenne* and white clover *Trifolium repens*. Such grasslands are referred to as ‘improved’ – as in improved for agriculture – so botanists and conservationists are always on the look-out for ‘unimproved’ grasslands (or even ‘semi-improved’ grasslands) as these will normally have more wildflowers and other wildlife. Unfortunately, though, conservationists have confused matters by coming up with a load of other terms for the types of grassland they’re interested in. So, for example, the old UK Biodiversity Action Plan used to include lowland pastures within its definition of ‘lowland meadows’. But, to add further confusion, the Biodiversity Action Plan also included separate categories for upland meadows such as those found in the Yorkshire Dales, and there was ‘lowland calcareous grassland’, on thin, well-drained soils over chalk and limestone (and which would usually be pastures though sometimes meadows), and ‘lowland acid grassland’, which was again well-drained but on acidic, nutrient-poor soils. These distinctions made sense for the purposes of planning conservation work, as each particular category had its own problems and demanded its own approach, but it does all add to the number of different ways of slicing the grassland cake.

If one thing is worth emphasising, though, it’s that a wildflower meadow (whether managed by grazing or mowing) is a type of grassland made up of perennial plants. A piece of colourful, flower-rich habitat which is made up of annual plants – like many so-called ‘pictorial meadows’ – is not really a meadow at all in an ecological (or farming) sense, and on this basis, neither should arable fields be thought of as wildflower meadows, no matter how beautiful they may be. In the end, of course, everyone is free to use terms like ‘wildflower meadow’ in any way they like. Language changes, after all, and insisting on the strict use of words like ‘meadow’ looks like going against the tide of history: when I searched online for images of ‘wildflower meadows’, 35 of the first 50 pictures were not meadows at all in the strict sense, but patches or beds or whole fields dominated by annual plants. Even the website for the Royal Botanic Gardens, Kew illustrated a page on ‘why meadows matter’ with an image of annual cornfield weeds.

But being pedantic about definitions is essential if you want to understand what another person might mean when they talk about a wildflower meadow. When you hear a figure quoted like ‘97% of wildflower meadows have been lost over the last hundred years’, then you need to understand that what is being referred to is flower-rich, unimproved, permanent grasslands. Knowing this difference is important because a long-established wildflower meadow is not easy to replace or recreate: in their own way, these grasslands are every bit as complex as woodlands in the relationships between plants, soil-living organisms, geology, history and management of the land they occupy, and indeed between the plants and each other. A sown pictorial meadow may be stunningly beautiful and may attract lots of bees – and jolly good, too, if it does – but no amount can make up for the loss of rich, intricate, real wildflower grassland.

When Plantlife took on Ranscombe Farm as a reserve in 2005, the site was almost entirely woodland or arable. Grassland was limited to one field (known as the Knock) which had been sown to grass in the 1980s and was used by the farmer for grazing cattle, plus two tiny bits of old, species-rich chalk grassland which had become more-or-less part of the surrounding woodland. One of these small grassland areas was tucked into the woodland at Mill Hill, and the other was squeezed between the edge of Cobham Woods and the arable of Kitchen Field.

The grassland on the edge of Kitchen Field had all but disappeared, but still held a smattering of chalk grassland plants which included the cucumber-scented salad burnet *Poterium sanguisorba*, yellow-flowered common bird's-foot-trefoil *Lotus corniculatus*, wild strawberry *Fragaria vesca* with its tiny and very tempting fruits, and man orchid. This small patch was rapidly vanishing under trees and scrub, and had probably never been a large area of grassland. But it had certainly been more open in the past: this is evidenced by several oak trees which have sown themselves (or been sown by jays *Garrulus glandarius*) in the grassland, and which have developed the wonky shapes and low branches that only form where a tree has grown in the sunlight and has not been forced up tall and straight by the shade of surrounding woodland. These oaks are still there, but much of the other scrub has since been deliberately cleared to restore as much as possible of the lost grassland.

The grassland at Mill Hill had a different history, being the remaining fragment of what had once been a much larger area of pasture, and it tells exactly the story of what happened to so much of the flower-rich grassland of the North Downs. As described in a previous chapter, the woodland at Mill Hill has developed over the last two centuries or so, on land that was previously open pasture or arable land. A combination of tree-planting and abandonment meant that by the early 21st Century, what had originally been more than 6.5 acres (more than 2.5 hectares) of pasture had shrunk to less than half an acre: a loss some 94% over around 200 years. Fortunately – and possibly due to rabbits keeping up some grazing during the middle of the 20th Century, or the stray sparks of steam trains on the adjacent railway line – what was left still held a good range of the downland plants I've already mentioned, such as common rock-rose, greater knapweed, horseshoe vetch and harebell, though of course this was just further evidence of the value of the grassland that had been lost. This tiny, remaining fragment had for some years been kept open by the sterling efforts of local conservation volunteers from the Cuxton Countryside Group, but young trees were continuing to grow and shade the remaining grassland.

This sort of loss, as grassland is abandoned, can creep up on you. Without management by cutting or grazing, the grass grows tall and rank and flowers disappear (as mentioned earlier, the lovely dropwort, with its frothy white flowerheads, was perhaps present into the early 1980s, but is

Clustered bellflower is rare in Kent, being found only on chalk grassland in a small area of the North Downs to the west of the River Medway.
Photograph: R I Moyse/Plantlife.

there no longer) and as the trees and bushes grow, the grassland itself slowly vanishes. The relatively slow pace of this change – it can take decades, rather than years – means that it may go unnoticed by anyone who regularly visits the site as you are only aware of slight changes between visits. But compare before-and-after photos, and the difference may be stark: at Mill Hill, aerial imagery shows that the grassland lost more than half its area between the 1940s and the 2010s. Go back further, and there is an old, postcard-sized photograph from the very early 20th Century which shows a wide, grassy, grazed area – there is a line of wooden hurdles to keep in sheep – surrounded by a narrow band of low scrub; a hundred years on, all that was left was a tiny island of grassland surrounded by tall trees.

If chalk grassland is allowed to vanish completely, it can be very hard to restore, as few grassland plants persist as seeds in the soil, and the soil itself is changed further by the accumulation of years of leaf-litter which makes it richer, deeper, and less able to support the plants that make chalk downland so interesting and special. If you want to rescue and restore chalk grassland that is vanishing in this way, it does mean cutting down trees and clearing scrub, and this sort of rapid change to the landscape – reversing decades of gradual grassland loss in just a year or two – can look sudden and brutal and is not always a popular move. But this is what was necessary and was done at Mill Hill, thanks to funding received from Natural England: trees and scrub that had grown up over the last 60 to 80 years were cleared, and machinery brought in to grind out the cut stumps. Now the grassland at Mill Hill probably covers about the same area as it did in 1940, and, though this is still only a little over 10% of the area of grassland that once existed here, it has provided more space for the chalk grassland plants to spread and increase their populations. Five years or so after the work was done, the newly restored area didn't completely resemble the old, original grassland, but it was filled with flowers, and man orchid, clustered bellflower and other chalk grassland plants had already moved in, and in good numbers.

These small but very nicely formed bits of grassland might well have been it for Ranscombe Farm if Plantlife hadn't been able to secure substantial funding from the National Lottery in 2010. This allowed them to add to the reserve a block of nearly 60 acres (23 hectares) of grassland at Brockles Field, in the south-west corner of the site. This big bit of land is steeply sloping for much of its area and faces south, so it is perhaps surprising that for much of the 20th Century it was put to the plough as arable land. Prior to this it had been pasture (the old tithe schedules show that at least part of it was pasture in the 1840s), but it only returned to grassland in the late 1980s, when compulsory set-aside was introduced. Set-aside was intended to deal with the surpluses of various agricultural

Wild liquorice is abundant at Mill Hill and in the grassland at Brockles Field, but unfortunately is not edible. Photograph: R I Moyse/Plantlife.

products which had been created as a result of European agricultural policies: in what was as a very simple precursor to modern agri-environmental measures, farmers were simply required to take a certain amount of land out of production.

This act, together with perhaps occasional rough cutting of the resulting vegetation, seems to have been enough to turn Brockles Field into grassland – though with a lot of scrub mixed in. Plantlife's work since acquiring the land has really been limited to just treating the field as grassland and letting the plants sort themselves out: initial cutting and clearing of the very dense, unmanaged, scrubby vegetation has been followed by seasonal grazing. What did not happen until very recently was any deliberate attempt to introduced seeds of grassland plants. As a result, we've been able to see how good plants are at DIY habitat-creation: what could be seen as a local, small-scale version of the way plants originally came together to form the first permanent grassland habitats.

A survey in summer 2019 of the flowering plants in Brockles Field found more than forty species, and this excluded the various grasses and woody plants (as well as some others, like grass vetchling *Lathyrus nissolia*, which, because of its grass-like leaves, is almost impossible to spot once it has finished flowering). Common grassland plants including wild basil *Clinopodium vulgare* (sadly, no good for cooking, despite the name), wild marjoram *Origanum vulgare* (usable as a herb, and in fact the same as oregano), hedge bedstraw *Galium album*, and wild carrot *Daucus carota* were amongst the most abundant plants recorded. These were accompanied by some more interesting species which included wild liquorice *Astragalus glycyphyllos* (another culinary disappointment: it only looks like the liquorice plant used to make the sweets, and doesn't have the flavour), the oddly-named common gromwell *Lithospermum officinale* (which produces small, pearly seeds which are as hard as stone) and the yellowy-brown, parasitic plant common broomrape *Orobanche minor*. There were even a number of plants that are typical of good, flower-rich chalk grassland such as yellow-wort, the delicate, white-flowered fairy flax *Linum catharticum* (also known by the less pretty but more practical and medicinal name of purging flax) and hairy violet *Viola hirta*, the typical violet of the open chalk downland. Together, these plants make for a diverse and very flowery grassland which supports a wide range of other wildlife. Just to take a few species also found by the survey, common bird's-foot-trefoil is used by common blue butterflies

With its grass-like foliage, grass vetchling is almost impossible to spot till its pink flowers appear.
Photograph: R I Moyse/Plantlife.

The small and moth-like grizzled skipper butterfly lays its eggs on wild strawberry.

Polyommatus icarus, dingy skipper *Erynnis tages* and the elusive six-belted clearwing moth *Bembecia ichneumoniformis*; grizzled skipper butterflies *Pyrgus malvae* are there because of the presence of their favourite foodplant, wild strawberry; and red clover *Trifolium pratense* and red bartsia *Odontites vernus* are favoured by one of Britain's most threatened bumblebees, the brown-banded carder bee *Bombus humilis*.

Repeating the survey in 2021 showed that three decades into its existence, the grassland at Brockles Field is still assembling itself. Several new plant species had arrived in the two years between the surveys: now present were greater knapweed, salad burnet, restharrow *Ononis repens* (a tough, pink-flowered member of the pea family), and yellow-rattle *Rhinanthus minor*. Interestingly, this last plant, which is a semi-parasite known for its ability to restrict the growth of grasses to the benefit of other species (and so a favourite of those aiming to create new wildflower meadows) is not known from anywhere else in Ranscombe Farm Reserve. Meanwhile, those species already present continued to jostle around. By looking at the whole field in 100m by 100m blocks, it was possible to see that, on average, the number of species in each individual block had increased between 2019 and 2021, mainly as plants spread around and became more evenly distributed across the field. The field also seemed to be changing character, with the arrival and spread of plants seen as typical of chalk grassland: large thyme, dwarf thistle, small scabious *Scabiosa columbaria* and wild liquorice were all more widespread in 2021 than they were in 2019.

The development of Brockles Field as a grassland has been astonishing to watch. The plants are doing very much the same thing as those plants that combined to form the first permanent farmed grasslands: just by treating the land as grassland, it was possible to create the opportunity for plants to come together and form a community which is diverse and rich in species and which is starting to resemble Ranscombe Farm's remaining fragments of chalk grassland. This is what can happen, given the right circumstance, if you let wildflowers be wild and do not rely on deliberate sowing to try and create the kind of habitat you think ought to be in a particular place. What is difficult to guess is how all these plants made their way to Brockles Field in the first place. Most are local and occur elsewhere on the reserve or nearby, and a few undoubtedly survived on the very edges of the field through the period when Brockles was routinely ploughed for arable cropping. Some of the new arrivals – a few – have wind-blown seeds, and this is probably how plants like bee orchid, pyramidal orchid and hoary ragwort *Jacobaea erucifolia* arrived. Others, like agrimony *Agrimonia eupatoria*, have seeds designed to latch onto animal fur (and people's clothes). But most have no obvious adaptations or designs that help them

Yellow-rattle in Brockles Field - it probably only arrived in around 2020, and has spread rapidly since. Where it came from remains a mystery.

spread: yet they can and they do, and perhaps this is all we really need to know.

We do get a couple of little hints from Brockles Field about how plants colonise a new space. Common eyebright *Euphrasia nemorosa* is another small, semi-parasitic plant that latches onto the roots of grasses and other plants; its branched stems bear small, toothed leaves and white, two-lipped flowers marked with violet and yellow. It seems to have arrived in Brockles via some deliberately created bare ground. As an experiment, Plantlife had, in 2011, created five shallow 'scrapes' in Brockles Field by removing the soil down, more-or-less, to the underlying chalk. The aim was to see how this might encourage development of chalk grassland vegetation. What happened first is that plants from the long-vanished arable suddenly reasserted themselves, with the reappearance of prickly poppy, rough poppy and other annuals from seed which had been buried for at least 25 years. This was followed by the gradual arrival into the bare, chalky ground of chalk-loving plants already present elsewhere in the wider field, such as the lovely carline thistle *Carlina vulgaris*, a biennial plant with brown flowerheads that look for all the world like dried flowers. Then, after two or three years, a small patch of common eyebrights turned up one of the scrapes. Eyebright seeds are small and may have been wind-blown, and, looking at locations of nearby populations, it seems likely that they travelled a couple of miles and perhaps more in order to reach Brockles Field. But however they got there, the few plants that established in that first year were enough to form the bridgehead for a coming invasion: in the second and third years after their arrival, plants spread out into the other scrapes, and then they started turning up in the wider grassland. They are now widespread throughout the field.

Building a bridgehead may be important. The single cowslip *Primula veris* which turned up in Brockles Field one year – again from no obvious local source – had gone the following year, and it has not been seen since. So it seems that plants need two things to happen if they are to establish a population on a new site. First, the seed needs to land in a spot which is right for germination and for the establishment of an adult plant. Second, seeds need to arrive and plants to establish in sufficient numbers to form a stable population: just one plant is unlikely to be enough to found a new population, as it may need – for example – to be pollinated by another individual and it will most definitely be more vulnerable to sudden, random disappearance than would a decently sized population.

One of the things that is noticeable from spending so much time staring at the vegetation on a single nature reserve is a sort of background churn in the plant species present, as new plants arrive, spend a year or two trying to make a go of it, then vanish once more. This seems to happen across many different habitats, not just in grassland. In some cases, the causes are clear or at least can be guessed at, as when sand was brought in and stockpiled by UK Power Networks as part of their work burying power cables. Hare's-tail grass *Lagurus ovatus* and grass-leaved orache *Atriplex littoralis*, two plants of sandy, coastal habitat, had clearly arrived as seed buried in the sand; they popped up only to vanish just as quickly when the sand was all used up. Hoary cinquefoil *Potentilla argentea* – a low-growing, yellow-flowered plant which is rare in Kent and considered Near Threatened nationally – turned up in some numbers in the woodland at Merrals Shaw, where it appeared along tracks which had very recently been used by forestry contractors working on the reserve. It is a plant of free-draining sands and gravels, so its appearance on damp soil over clay-with-flints was a surprise: it is perhaps most likely that it arrived by chance on the wheels of the forestry vehicles. Despite considerable efforts to manage the trackside habitat to benefit hoary cinquefoil, it vanished within a couple of summers, adding weight to the idea that it came with the contractors' vehicles, had a go at setting up a population, and ultimately found the habitat unsuitable. Another plant that may have been unwittingly carried in was hound's-tongue *Cynoglossum officinale*, a handsome, purple-flowered plant which, strangely, smells quite powerfully of mice. A single plant of hound's-tongue appeared one year on the edge of a field, and may have come in stuck to a visitor's clothes or even their dog: the plant is found in good numbers on other nature reserves not far from Ranscombe Farm, and its seeds, which are covered in short hooks, are very adept at sticking to fabric or fur.

In the case of some other plants, how they ever arrived at Ranscombe is a complete mystery. Floating sweet-grass *Glyceria fluitans*, usually found in marshes and pond-edges, turned up once in a shady woodland ride, and grew enough to put up a spike of flowers, but was gone almost straightaway. A small temporary pond in Cobham Woods was, for one summer, home to brackish water-crowfoot *Ranunculus baudotii*, a kind of white-flowered wetland buttercup which, as its name suggests, is more usually found in coastal ditches and lagoons. Chicory *Cichorium intybus*, with its ragged, sky-blue flowers, appeared in the arable at Kitchen Field one year. In all these cases, it seems that getting to Ranscombe wasn't the problem – what let the plants down was either arriving in numbers too small to set up a viable population, or that the habitat just wasn't what they needed to survive in the long term. Yet none of these three species is obviously equipped for long-distance travel.

The ability of different plant species to find their way to a new site is fascinating, whether their attempts are successful or otherwise. Just as interesting is the apparent inability of some species to travel at any detectable speed. There is an excellent paper by Joe Costley in the Conservation Evidence journal that addresses this issue. It is based on a study at Plantlife's Joan's Hill Farm reserve in Herefordshire, where the main part of the reserve consists of a cluster of small hay meadows which are separated from each other by hedgerows. Some of these meadows had been grassland for at least 170 years when the study started in 2002, and were especially rich in grassland plants, while others had more recently been managed as arable and were consequently much poorer: it was one of these species-poor fields which was selected for the study. Over the next 11 years, the meadows were all cut for hay each year, and grazed by cattle after cutting, with the cattle being able to wander freely between the various meadows while they were on the site. At the end of the 11-year period, the number of plant species in the study field had increased from 24 to 34, and the field as a whole was more diverse. However, despite the close spacing of the fields, and the movement of cattle and farm machinery between them, it was found that there were some plants which were still only present in those fields which had been managed as meadows for many decades. Pepper-saxifrage *Silaum silaus*, common milkwort *Polygala vulgaris* and betony *Betonica officinalis* were mainly confined to those fields thought to have been grassland for at least 88 years; and established colonies of green-winged orchid *Anacamptis morio* and dyer's greenweed *Genista tinctoria* were only found in fields believed to have been grassland for at least 170 years.

Ranscombe Farm Reserve has not had the advantage of livestock wandering freely between patches of grassland, though, as I've suggested in the chapter on orchids, man orchids may have arrived in Brockles Field as a result of moving livestock from the nearby grassland in Kitchen Field. It also seems likely that cattle may play a role in the spread of plants from place to place within Brockles. Seed can easily get trapped in mud on the feet and hair of domestic and wild animals, and studies in Europe have shown that large mammals may be responsible for moving a lot of plants around the landscape. One scientific paper by European researchers described how they carefully searched the feet and fur of 25 roe deer *Capreolus capreolus* and nine wild boars *Sus scrofa* that had been shot by hunters, and were able to find the seeds of 55 different plant species of grassland and disturbed ground. Ranscombe Farm does not have wild boar, but it does have fallow deer *Dama dama* and it is also visited by lots of walkers, often accompanied by dogs: between them, people and animals must pick up a lot of seeds in the mud on their feet. Many walkers and their dogs pass through the patch of old, species-rich chalk grassland at Mill Hill every day, with a large proportion continuing up the footpath into Brockles Field. The distance is just a few hundred metres, yet plants like clustered bellflower, horseshoe vetch, dwarf thistle and burnet-saxifrage *Pimpinella saxifraga* – all present in good numbers at Mill Hill – don't seem to be able to successfully make the journey. This suggests that either the conditions in Brockles Field – the soil, the existing vegetation – are not right for these plants to establish, or that these species are amongst those that have no ability to drift on the wind or hitch a lift on a passing mammal.

Clearly, every one of these plants must have, or must once have had, some mechanism that allows it to travel through the landscape: clustered bellflower, for instance, is found throughout England and up into Scotland on chalk and limestone soils and so must have got around somehow. Yet at Ranscombe Farm Reserve, it seems to stick stubbornly within the little patch of grassland at Mill Hill. So perhaps its method of travel is no longer effective in the modern countryside. Or perhaps it is so slow that two or three decades is not enough time. Either way, this situation puts Plantlife, as managers of the reserve, in something of a dilemma. Do you continue to give the plants time to make their own way from the established chalk grassland and up into Brockles Field? Or do you take the position that the restoration of chalk grassland – with all its component species – is such an urgent conservation priority that some deliberate human intervention is needed? The choice is not so obvious as it may first appear: it may seem quick and easy to move plants about and sow seed, but in doing so you can lose sight of plants as truly wild beings, and of the need to understand and restore the natural systems that once allowed plants to spread through the landscape.

In this case, it was felt that as so many plant species were apparently unable to travel just a few hundred yards in more than 35 years, then a helping hand was justified, and the simplest method for doing that was by moving green hay. If nothing else, this would go some way to answering the question of whether the barrier to clustered bellflower and its colleagues setting up shop in Brockles Field was an inability to travel or the unsuitability of Brockles as a new home. Green hay provides a kind of quick-and-dirty way of introducing wildflower seeds to a new site. You simply cut an existing area of grassland once most of the flowers have set seed but before all the seed has dropped – so generally mid- to late summer– and rapidly move and spread the cut material onto the new patch. It needs the new patch to be bare soil or to at least have very short and open vegetation, so that seeds can reach the soil and germinate. The cut and spread material is left to shrivel and dry and to drop any seed it is carrying. Then you wait and see if it has worked – grassland perennials need at least a couple of years to get to flowering size, so a bit of patience is required.

From the point of view of plant conservation, the use of green hay has the advantage that the seed being introduced reflects the mix of local plants in a way that a packet of seeds cannot. But, as with any deliberate introduction of plants to a new site, green hay does have its disadvantages. In particular, you lose information about the plants you are dealing with. If we find clustered bellflower or horseshoe vetch in Brockles Field now, we'll never know whether they got there themselves or were brought in the green hay. For this reason, when green hay was spread in Brockles Field, it was done in a restricted and mapped area: this meant that a plant appearing in that area before it was found elsewhere in the field was most likely to have arrived in the hay. As yet, results are still pending.[1]

Although widespread in Britain, clustered bellflower is a rare plant in Kent, with a very localised distribution on a tiny stretch of the North Downs covering just a few square miles to the south-west of Ranscombe Farm, and many of its known sites are on privately-owned land where appropriate management cannot always be guaranteed. So it might be argued that the case for getting it into Brockles Field, where it could potentially form a very large population, outweighs the loss of any information about how the plant might be able to move without human assistance. Perhaps so, but it's also the case that if we want wild plants not just to be limited to nature reserves, then we do need to know what they need in order to be able to spread themselves through the wider landscape. And, at the very least, we need to understand which species might not be able to do it without deliberate human intervention. A case in point is meadow clary, mentioned in a previous chapter and one of the grassland plants of most interest and concern at Ranscombe Farm Reserve.

Meadow clary is a large and quite muscly-looking kind of sage. It is a perennial which dies back completely in winter, but which in spring pushes

[1] No longer. In summer 2022, after this was written, three clustered bellflower plants were found, in flower, in an area spread with hay some five years previously.

up rosettes of wrinkly leaves that are followed by tall, branched heads of big, deep-blue flowers. In Britain, it is at the north-western extent of its European range, with native populations as far west as Gloucestershire, but it is nowhere common in this country. In Kent, there are just three remaining sites, with Ranscombe Farm being the stand-out site because of its long history following the plant's original discovery back in the late 1600s.

Meadow clary flowering on the edge of Cobham Woods.
Photograph: R I Moyse/Plantlife.

Meadow clary has a reputation as a tricky plant to get to grow and spread, but it is also a plant that can encourage a blasé attitude. Adult plants are long-lived and flower repeatedly, and will also branch below ground to produce new rosettes, meaning that what looks like a cluster of small plants can in reality be just one big one. So a population may look large and healthy when in fact it consists of no more than a small number of ageing individuals that are failing to reproduce: in such cases, you can end up with what is sometimes called an 'extinction debt', where everything looks OK, but the population is ultimately doomed to vanish when the plants all finally die of old age. The key to preventing this and maintaining a healthy and viable population, as well as getting that population to grow and spread, is to have regular recruitment of new generations of seedlings. At Ranscombe Farm Reserve, some serious work on improving the health and viability of the meadow clary population started in 2012. At this time, despite a continuous presence at Ranscombe Farm of more than 300 years (and sterling work by wardens Helen Morley and Ben Sweeney to get the numbers up from an initial low of just six plants) the entire population consisted of no more than eleven meadow clary rosettes which had put up heads of flowers, and a further ten rosettes which looked like young plants but which turned out to actually be fully grown plants that simply had not flowered. In some cases, lack of flowering was due to plants being right on a footpath and getting walked on daily, something fairly easily solved with a little bit of fencing to stop the trampling: the fence went in and the plants flowered the following summer.

Meadow clary. Left: a first-year seedling, still with its D-shaped seed-leaves. Right: rosettes of immature plants. At Ranscombe, plants have taken five years to go from germination to first flowering. Photographs: R I Moyse/Plantlife.

The next step was to try and get the plants reproducing, which surprisingly turned out to be relatively simple though perhaps not immediately obvious. Meadow clary flowers are very attractive to bumblebees, and most flowers will set seed. Most of the seed also seems keen to germinate, but only where it drops onto bare ground, and the tiny, new seedlings need light to grow: a seedling that gets swamped by grass or other plants is doomed. What we discovered was that these conditions could be provided by spending an hour or two with hand-tools scraping off the surface vegetation around the patches of flowering plants, so that there was bare ground around each of them for a distance of up to a metre or two. As one of the people wielding the tools, I found this to be one of those very satisfying jobs where an idea – admittedly one culled from reading the published literature about the plant and making observations elsewhere – actually worked straight away. The seedlings, with their little, D-shaped seed-leaves, turned out to be easy to identify (at least if you got down on your hands and knees) and in the first autumn more than 120 of them were found, making the population instantly seven times as big as it had been just months before.

Of course, it is tough being a seedling, and not all survived, but with a bit more creation of bare ground every two or three years, and more seedlings appearing every year, the population in 2021 stood at a total of 95 plants. It is a major increase and means that the population is now far more secure, but it took right up to 2021 to get a serious increase in the number of grown-up plants which were big enough to put up some flowers. Because meadow clary – at least in the wild – is a seriously slow-growing plant, taking as long to mature as some orchids. The fastest growing plants at Ranscombe Farm Reserve – those in full sunlight – took five years from germination to first flowering, and where plants are in partial shade, it has taken eight years. As a result, in 2021 there were 35 flowering plants, compared to 20 back in 2012, so that while the overall population is more than four times what it was, the number of flowering individuals has not even doubled. The key thing, though, is that there are 60 more plants of all ages, from seedlings upwards, that are coming up through the ranks ready to form future generations of flowering plants.

It is telling that out of the four little, separate groups of plants, the patch of meadow clary that consistently produces the most seedlings is not the one with the most plants. The most numerous patch only ever produces a few seedlings, and these often disappear rather quickly. I suspect that what this tells us is that this big patch, which looks to the eye like 14 separate flowering plants, may in reality just be one or two individuals that have branched and split below ground and which are all so closely related that they suffer what is called 'inbreeding depression', where any seeds and seedlings are puny and unfit. It's another lesson about not assuming everything is OK because you can see a good number of apparently healthy plants: this particular cluster may actually be doomed in the long run unless it gets an injection of new genes from one of the other patches.

There has been real success in getting the numbers of meadow clary plants to increase, but have they spread? The field which currently holds the whole Ranscombe Farm population has been fenced and is grazed annually by cattle in order to create another 7 or 8 hectares (about 20 acres) of grassland on what was previously arable, in the hope that it will eventually hold a major meadow clary population. Cattle are great at tramping round in the wet winter weather, opening up the soil and hopefully avoiding the need for further input by humans with spades and mattocks, and, with any luck, the mud the animals cart round on their hooves will sometimes hold a few meadow clary seeds. Certainly the plant needs help in this respect, as the seedlings found so far have all been within a metre of the parent plant. Only time will tell whether the cattle have done the job, when any new plants that might be out there, unseen amongst the long grass, finally raise a flag in the form of a big spike of smart blue flowers.

Chapter 8 - The gene genies

It may seem unnecessarily difficult and time-consuming to create new wildflower meadows by letting livestock and the plants themselves do the work. Why not plough and reseed with a wildflower mix, especially given the urgent need to reverse the massive historic losses of flower-rich grassland in this country? The answer, I think, depends on what you think plants are there for. Creating a wildflower meadow from seed is quick (though expensive) and undoubtedly effective, but you get a meadow with a form and composition that has been decided by humans. If you see plants as being not always at the beck and call of people, but as creatures which are every bit as wild as birds or butterflies, then it is clearly essential to interfere as little as possible. The job of people then becomes to provide plants with opportunities and to remove the barriers that prevent them from being properly wild. This may sometimes mean ploughing and seeding, or adding seed or plants to existing grassland, but only after careful thought and never as the first option. As shown by Plantlife's Joan's Hill Farm reserve, or the experience with clustered bellflower at Ranscombe Farm, some plants find it hugely difficult to move around the modern countryside. But there are plenty of plants, like the common eyebright that came to Brockles Field, that are well able to colonise the wider landscape if they are given the right opportunities. The history of permanent grasslands in Britain, and the more recent history of Brockles Field at Ranscombe Farm, show that plants have an amazing ability to come together to form new communities in response to the way land is treated. With the current, rapid, global changes to the environment, we need wild habitats to be able to adapt, change and evolve. We may even need individual wild species to adapt and evolve. Fortunately, it turns out that plants are very good at this too.

The common eyebrights that came and made their home in Brockles Field are a good example, being, as they are, part of a group of plants that seem to be in a process of active evolution. Eyebrights are short, annual plants with small, toothed leaves. The stems bear pretty, two-lipped flowers which are very pale or even white in colour but are variously marked with violet – especially along the veins – and with yellow patches. It is the colouring of the flower which originally gave the plant its name, as the early herbalists saw the purple veins and yellow flushing as being similar to the features of a diseased eye: as a result, the plant had a long history as eye medicine. Using the appearance of a plant to determine its usefulness was known as the 'doctrine of signatures': it was assumed that divine

Common eyebright - one of the many plants growing in the bare, chalky 'scrapes' in Brockles Field. Photograph: R I Moyse/Plantlife.

providence had indicated, through physical signs, the various medicinal uses to which plants might be put.

I did mention previously that eyebrights are semi-parasitic: that is, they obtain some nutrients from other plants, while still being able to use photosynthesis to feed themselves. That turns out only to be nearly true. The authoritative book on eyebrights in Britain, published by the BSBI, points out that they are able to survive without parasitising other plants, though free-living plants tend to be smaller. To me, this makes them even more fascinating. Plant parasitism is such a weird thing in itself – it seems such an animal-like behaviour – but it is positively spooky to have a plant that can switch from providing for itself to draining food and water from its neighbour. And eyebrights are not fussy, being able to latch onto a range of species from different plant families or even onto each other. Give them fangs and get them up and moving, and eyebrights would be standard horror-film fare. The weirdest thing about eyebrights, though, is the way they completely mess with our idea of what a species is.

We tend to think of any species – whether plant or animal – as being a distinct entity that is clearly and visibly different to any other, even if those distinctions are quite subtle. Once upon a time, we also used to think of a species as consisting of a group of individuals which were able to breed successfully with each other, but which could not breed with members of any other species. However, even my old Penguin Dictionary of Biology – the 1973 edition, from my schooldays – recognises that this idea of defining a species by this 'reproductive isolation' doesn't always work and that a consequence of this is that it takes an expert to decide what constitutes a single species and what doesn't. The experts used to work by carefully examining and measuring physical differences between plants to determine whether groups of similar-looking individuals should be split into separate species or grouped together as one, but nowadays, all the cutting-edge stuff is based on DNA analysis. Papers are published which use this sort of work to build family trees showing just how closely related different plants actually are, based on their genes – though even then it needs expert opinion to decide whether the differences between two groups of plants are sufficient for them to be considered as separate species. And even this has its limits: it is famously the case that the two British species of butterfly-orchid can easily be distinguished from each other when in flower (and just using the naked eye) but cannot be separated on the basis of their DNA – or, at least, not yet. With plants, defining a species can be a fraught process.

In an earlier chapter, I mentioned the tendency of orchids to baffle botanists by refusing to fit neatly into our ideas of what a species is or what it should be, so that some authors list 56 orchid species in Britain and Ireland, while for others there are a good deal fewer. The marsh-orchids and some of the helleborines are especially troublesome, and the marsh-orchids in particular are well-known for the freedom with which different species hybridise with each other. But eyebrights take this confusion to a whole different level. The BSBI handbook lists 21 species of eyebrights for Britain and Ireland but a massive 71 different known hybrids between these species. The handbook's authors point out that hybrids can be found for pretty much any two species that happen to be growing in the same place. Not only that, but the hybrids are usually fertile and can themselves hybridise with other hybrid plants or with either parent. As hybrids tend to show characteristics somewhere between each of their immediate parents, this can produce something called a 'hybrid swarm', where there is a large crowd of plants showing every degree of variation between each of the original, founding species. As a result, it is very difficult both to determine where an individual species starts and ends – what is its normal range of leaf-shape, flower-size, hairiness or other features – and to confidently identify any plant to a particular species. In fact, the handbook recommends that the identification of any eyebright to species should be based on collecting at least six specimens from a population and measuring a wide range of features. At Ranscombe Farm Reserve, we found, by collecting a set of specimens, that our plants were almost certainly the species called, in English, common eyebrightt, which makes things pretty straightforward, at least for now. But you do not have to go far – in fact just to some bits of chalk grassland within a few miles of Ranscombe Farm – to find big swarms of eyebright hybrids. And Ranscombe Farm does have its own hybrid swarms to deal with, but, in this case, it does not involve eyebrights.

There is a pretty little cornfield annual called the field pansy, which I mentioned in an earlier chapter. It is a weedy species that seems to be one of the survivors in the modern farmed landscape, not least because of its ability to pop up and flower in late summer after the crop has been harvested, and so avoid being shaded by the crop or being sprayed with herbicide. In general, the plants get to maybe 15cm (six inches) or so tall and produce a series of small, creamy-white flowers which are similar in shape to the flowers of violets (violets and pansies are very closely related, all being in the genus *Viola*). Look amongst the autumn stubble of even the most intensively farmed field, and there's a good chance you'll find them, and it is always a little moment of pleasure when you do. At Ranscombe Farm Reserve, they are plentiful, so when one of the reserve's wildlife survey volunteers, Chris Atkinson, sent me a photo of a white-flowered pansy he had spotted during a winter visit, I immediately assumed that it was a field pansy. It was only when I went and looked for myself that I realised I was wrong: field pansies have flowers with very small petals – so small, in fact, that the sepals (which enclose the developing flower and open with it) are easily visible when the flower is open. The plants which Chris had seen had much larger petals which hid the sepals, and when I looked more closely, I found there was a corner of a field where there were hundreds of pansy plants, all superficially similar with creamy-white flowers, but with a range of flowers sizes from a couple of centimetres (nearly an inch) down to around half this height. Measurements of flowers and examination of the leafy parts of the plants showed that this was a swarm of hybrids between field pansy and its near-relative wild pansy *Viola tricolor*: the two species had hybridised, and those hybrids had crossed with other hybrids and with their parents. This had produced a mass of plants which at one end of the scale looked very like field pansies and at the other looked very like wild pansies, but which probably all shared some genes from each of the original parent species.

One of the mass of hybrid pansies *Viola* x *contempta*, with features intermediate between field pansy and wild pansy. Photograph: R I Moyse/Plantlife.

Even more surprising is that we had never found proper, obvious wild pansies at Ranscombe Farm – and we would have been looking out for it, as it is an increasingly scarce plant, considered Near Threatened in both the British and English Red Lists. Wild pansy is a lovely plant, with flowers that can be pale yellow but which more often have strong blue or purple colouring; as a result it was brought into gardens long ago, where it is called heartsease. Wild pansy is also one of the parents of the garden pansy *Viola* x *wittrockiana*, which is a deliberate, human-created hybrid of three wild species: the British wild pansy and mountain pansy *Viola lutea*, and the Altaian pansy *Viola altaia* of Central Asia. Wild pansy presumably occurred on the reserve at one time (and I have found it nearby, around half a mile away as the crow flies) but has been totally absorbed into the mass of hybrids. So what do you call these plants? Officially, you call them *Viola* x *contempta*, the 'x' in the middle indicating that this name refers to a hybrid, but can you call some of them field pansies if that's what they look like, or wild pansies if they are at the other end of the spectrum? And if you cannot do this, then how do you know that any other field pansies you find are not really hybrids which just have too few wild pansy genes for it to really show? All of which makes you question what a species actually is, and whether the idea of a 'species' really has any meaning in this case.

Even oak trees aren't immune from this problem. The two native oaks are pedunculate (or English) oak and sessile oak. The common names of these two are all about the acorns – whether they are held on a long fruit stalk (technically called a peduncle) or without a long stalk and so more or less seated (sessile) on the twig – but there are other differences in the shapes of the leaves and even the shapes of the trees themselves, with sessile oaks tending to be taller and straighter. Pedunculate oak is more often found on deeper, richer soils, while sessile oak is a plant of shallower, more acidic soils, though they overlap and where they meet they hybridise – as they do at Ranscombe. The hybrids are fertile and so the same situation occurs as with the pansies: they can cross with each other and back-cross with either parent. As a result, there are oaks at Ranscombe Farm that look right for pedunculate oak, and a few that look right for sessile oak, but even on the sandiest soils in Cobham Woods, most trees have features that put them somewhere between the two.

Hybridisation is a genuine headache for botanists, being so common. There is even a published Hybrid Flora of the British Isles, which details more than 900 plant hybrids that have been found in the wild. While animals can, and do, hybridise in the wild, it is far less common than in plants and the resulting young are often sterile. But in plants, hybrids between closely-related species can often be fertile, as in the eyebrights, pansies and oaks discussed above. This means that hybridisation can provide a way for entirely new plant species to arise, and can be an important factor in plant evolution. This can even happen where the hybrid offspring are normally sterile, as plants have a clever trick up their sleeve. In an earlier chapter, I talked about trailing tormentil: how it is difficult to identify, and needs checking to see whether it has produced any seeds. Plants that look like trailing tormentil but produce very few or no seeds are probably one of the very similar, but sterile, hybrids which trailing tormentil can form with one or other of the closely related plants, creeping cinquefoil and tormentil. This subtle difference is less surprising when you find out the reason is that trailing tormentil originally started out itself as a hybrid between creeping cinquefoil and tormentil, but one which managed to become fertile (and so able to produce seeds) by doubling-up its chromosomes. This genetic trick is something plants seem to be able to pull off quite often, and which hardly ever happens in animals.

Here (roughly) is why chromosome-doubling works. The chromosomes, found in the nuclei of plant and animal cells, are the bodies containing the genes. In general, every plant or animal cell has two copies of each chromosome, one copy being inherited from each parent during sexual reproduction. To achieve this, the sex cells that meet and combine must each contain one copy of each chromosome. This is achieved during the production of the sex cells, when the chromosomes arrange themselves in their pairs; each pair then splits so that two sex cells are produced, each with just one chromosome from each pair. If sex cells from two different species come together, they may be able to fuse successfully to create a hybrid embryo, and this might be able to grow into an adult hybrid plant. But unless the hybrid's two parent are closely related, the differences between the two sets of chromosomes means that they will be unable arrange themselves in matching pairs – and that means it is not possible for the hybrid to successfully produce its own sex cells. Think of it like pairs of socks. If you have four pairs of short socks (so eight socks in total) with each pair a different colour – say, one pair each of red, blue, yellow and green – you can split each pair into two groups, each with one sock of each colour. You could then take another four pairs of socks of the same length and colours, and again split each pair in half – put the four socks from one pair with four socks from the other, and you would have no problem forming four pairs of matching socks.

But what if you try and 'hybridise' your socks by taking your four individual, short, coloured socks and combining them with three long, patterned socks, each taken from another pair? These socks might mix together fine in the sock drawer, but you would never be able to pair them up properly. This is what happens in some hybrid plants – the mix of different chromosomes is able to function until it comes to trying to produce sex cells, when the chromosomes are unable to form pairs and the system breaks down.

If a species or hybrid cannot create its own sex cells, it is never going to be able to reproduce and it will never be anything more than an evolutionary dead-end. In plants, however, it sometimes happens that, somewhere during sexual reproduction, the chromosomes double-up to twice the normal number – a double set from each parent – and so they can more easily pair up to produce sex cells that function fully. Sticking with the sock analogy, doubling up means you would have eight short, coloured socks – two of each colour – and six long, patterned socks, meaning that you could form seven good pairs of matching socks. The plant with a double set of chromosomes may not be able to breed with either of its parent species, again because of the mismatch in chromosome numbers. But because many plants are able to fertilise themselves – again something not generally available to animals – it may take just one chromosome-doubling event to set a new species on the road to success.

It appears that new species have arisen from hybridisation on lots of occasions during the long process of plant evolution. Another example is the little, white-flowered member of the cabbage family, wavy bitter-cress *Cardamine flexuosa* – this is an annual or short-lived perennial which is generally common in damp places and has been found at Ranscombe Farm. It has been shown that this started out as a hybrid between the taller wetland perennial large bitter-cress *Cardamine amara* and the little, weedy hairy bitter-cress *Cardamine hirsuta*, a very widespread species which can turn up all over the place, including in plant pots in gardens. Large bitter-cressand hairy bitter-cress both have 16 chromosomes, but in wavy bitter-cress, the number is 32 – a result of chromosomes doubling-up in the distant past. Wavy bitter-cress has some of the weedy characteristics of its hairy parent, and as a result is far more widespread than large bitter-cress, which is much more closely tied to wetland habitats; this could actually be seen as success for large bitter-cress genes, which have used hybridisation to increase their abundance across a much wider landscape.

It still turns out, though, that managing with a larger number of chromosomes can be a bit of a tricky job. If we are still thinking of chromosomes as pairs of socks, then the more you have, the greater the chance of some getting lost in the wash, and having multiple sets of chromosomes means plenty of chances for errors to occur. But plants have another potential mechanism to deal with this, which is to abandon sexual reproduction altogether, do away with the problems of chromosome pairing and division, and produce seeds which are direct clones of the parent. When this starts to happen in a population of plants, it means that all the subtle differences between individuals end up being preserved for generation after generation, producing a series of distinct, though still very similar, species – which are often called microspecies. Even though this might look in some ways like a dead-end, it is successful enough as a survival strategy to have been adopted by a number of different groups of plants.

The hairy lady's-mantle found in Cobham Woods is one of these species which takes this cloning approach, a process which goes by the technical term of 'apomixis' meaning 'without mingling'. Another example from Ranscombe Farm Reserve is goldilocks buttercup, which can be found in the little strip of ancient woodland at the top of Mill Hill. Popping up in April and May alongside other spring flowers, it is quite a distinctive plant. The deeply divided upper leaves are like spreading green fingers and make it look quite different from other common species of buttercup. But even

A selection of flowers of goldilocks buttercup. They may vary tremendously within a single plant.

more unusual are the strange flowers: though they are a straightforward buttercup-yellow, the individual petals vary greatly in size from maybe 7 or 8mm (around ⅓ inch) to tiny or absent altogether – and this can be all within a single flower. As a result, the flowers on just one plant can be all sorts of shapes and sizes, from the kind of even, rounded form you might expect on any buttercup, to an irregular mix of large and small petals, to barely there at all. According to the fourth edition of the New Flora of the British Isles, there could be as many as 200 microspecies of goldilocks buttercup in Britain, each distinct from the rest, but at the time of that book's publication only (only!) 58 had been described.

Apomixis can also be found in dandelions, most of which do not reproduce sexually, and where more than 200 microspecies are recognised in the British Isles, with probably more than 40 of these not found outside these islands. Brambles do it too, so that more than 300 microspecies are recognised for Britain and Ireland. So far, there has been no work to identify which of the various microspecies of dandelions and brambles are found at Ranscombe Farm Reserve; and as far as I am aware, no-one has yet looked at the goldilocks buttercups of Southern England, so it is by no means certain that Ranscombe Farm's particular species has ever been described. Though we are okay on hawkweeds – despite there being more than 400 microspecies in the British Isles, we know that the one which grows so handsomely along the Town Road in Cobham Woods is autumn hawkweed.

This weird plant thing of doubling-up the number of chromosomes doesn't just happen with hybrids, either. Sometimes a single species will do it. The lovely harebell has done it a number of times in its history. A few harebell plants still hang on in the chalk grassland at Ranscombe Farm, where they look so slight and delicate with their wire-thin stems and tissue-paper flowers: but don't believe the appearance – this is a plant tough enough to survive at over 1000m elevation in the Scottish mountains. The original harebell, with just a single set of chromosome pairs, does not seem to occur in Britain: in the south of Britain we have plants that have doubled up the original chromosome number, so that they have four of each chromosome instead of just two. Further to the north and west of Britain, plants of this type are joined by harebells with six of each chromosome, these plants having larger flowers and broader leaves and being sufficiently different to be identified as a separate subspecies. In mainland Europe, where there are many more related species, the situation is even more complicated, with some researchers struggling to find clear characteristics which can be used reliably to separate some of the apparent species. The authors of a paper on harebells in the Journal of Ecology put this problem down to on-going evolutionary processes: in other words, the plants are continuing to change and evolve as we watch them.

This continuing evolution is something we tend not to think of much. We have in our heads the idea that evolution is something that has happened in the past and which has left us with a planet-full of plants and animals which are distinct from each other and largely unchanging. We might accept that some species are sufficiently similar to each other that it takes care and close observation to tell them apart, but it is a bit of a leap to persuade ourselves that this may be because those species are in a process of active change. But evolution will always happen if a new combination of genes arises and turns out to be better suited to exploit the environment than any existing set of genes. Active evolution is almost inevitable in situations where new hybrids can arise, or where an existing species finds itself in a new situation with new environmental pressures or opportunities, or where we have changed the environment around a species.

The difficulties with identifying eyebrights, and the huge number of different hybrids between them, has been put down to this whole group being in a phase of active evolution. The same is said to be true of many orchids, including the group to which Ranscombe's common spotted-orchid belongs, the genus *Dactylorhiza* (the name means finger-root, as they have tubers with finger-like branches, rather than the egg-shaped tubers of many other orchid species). This group includes a lot of similar-looking species which frequently hybridise with each other to create all sorts of intermediate forms. Some of these hybrid orchids have stabilised over time into consistent forms which are considered species in their own right: one such is the narrow-leaved (or Pugsley's) marsh-orchid *Dactylorhiza traunsteinerioides*, which is found in parts of North Wales, Northern England, Scotland and Ireland, but nowhere else in the world. Presumably, its parent orchid species first got together somewhere in the British Isles soon after the last retreat of the glaciers, as they were making their way up from mainland Europe to colonise the newly available landscape.

Or perhaps it was early human activity that brought this orchid's parent species together, creating a novel environment where the two parents were able to meet in a way they might previously never have done. In the

Common bird's-foot-trefoil - a plant created by natural hybridisation between two other species. Photograph: R I Moyse/Plantlife.

previous chapter, I've referred to George Peterken's book on meadows and the idea that the plants that make up today's managed pastures and meadows originally occupied a range of very different wild habitats. This means that the advent of livestock farming in the Neolithic brought together plants that may previously have met only very rarely. Peterken suggests that one result may have been the evolution of common bird's-foot-trefoil. This very common grassland plant, with its yellow pea flowers and association with many grassland insects, is known to have hybrid origin: its original parents were most probably greater bird's-foot-trefoil *Lotus pedunculatus* and narrow-leaved bird's-foot-trefoil *Lotus tenuis*, plants which today are found in wet and dry grasslands respectively (greater bird's-foot-trefoil can be found in damper parts of Cobham Woods). If farming had never come to Europe, then common bird's-foot-trefoil may never have come to exist and wildflower meadows in Britain would look very different today, right down to the butterflies and other insects that use them. But there's another plant – a much, much rarer plant – that may well have its origins in farming and which has featured strongly in recent work at Ranscombe Farm Reserve, and that's interrupted brome *Bromus interruptus*.

Interrupted brome is an annual grass which germinates in autumn or sometimes in spring, sending up flowering shoots which – at a stretch and in tip-top conditions – can be a couple of feet (60cm) or more tall. Like all grasses, its little flowers are clumped together into spikelets, with a number of these spikelets making up each flowerhead. In interrupted brome, the spikelets are a dumpy, oval shape, and on very short stalks, making a compact head. Most of the spikelets are in a clump at the top, but there is often a smaller clump below, with a short gap separating the two: this 'interruption' in the flowerhead gives the plant the first part of its common name (and the second part of its almost comically similar scientific name). The bromes are a group of annual grasses with similar, more-or-less oval spikelets of flowers, though in many species, these are on long stalks. At Ranscombe Farm Reserve, for example, you might also see (amongst others) the very common soft-brome *Bromus hordaceus* or the taller rye brome, both with spikelets on stalks.

A flowerhead of interrupted brome.
Photograph: R I Moyse/Plantlife.

No-one knows exactly where interrupted brome came from. Despite the long history of botanical exploration in Britain, interrupted brome wasn't discovered until 1849 – when it was thought to be a variety of soft-brome – and it was not until the 1890s that it was formally recognised as a species in its own right. Over the following few decades, this newly discovered species was found to be widespread in Southern and Eastern England. However, it was never found outside England, except for a couple of records of possible accidental introductions to the Netherlands in the 1930s, and it is still not known from anywhere else in the world. Then it seemed to vanish as quickly as it had appeared: from 1936 onwards, just six sites were found for it, and by 1962, interrupted brome was known from just one location in Cambridgeshire. It was last seen there in 1972, and, though seed had been collected from this

Interrupted brome in the Southern Valley at Ranscombe.
Photograph : R I Moyse/Plantlife.

last population, this turned out not to be viable and the plant was thought to be extinct in Britain. As it had never been found outside Britain, it was therefore also globally extinct.

Then, a near-miracle happened. At a plant conference in 1979, Dr Phillip Morgans Smith, a lecturer at Edinburgh University, brought along some interrupted brome plants descended from seed that, unknown to others, he had collected in the early 1960s from that last known site in Cambridgeshire. This allowed a start to be made on a rescue operation, with seed from Dr Smith's plants being grown on at Edinburgh and Cambridge Botanic Gardens, at Paignton Zoo, and in other collections to 'bulk up' the total amount of available seed. Seed of interrupted brome now forms part of the collection in the high-tech Millennium Seed Bank, located at Wakehurst Place in Sussex and run by the Royal Botanic Gardens, Kew. Having a safe supply of plants and seed held in captivity, as I suppose you could call it, meant that attempts could be made to reintroduce it into the wild.

In 2015, Plantlife was approached by Dr Fred Rumsey, of the Natural History Museum in London, to ask about the possibility of using Ranscombe Farm Reserve as one of a small number of sites for the reintroduction of interrupted brome. Any reintroduction of any species –

whether plant or animals – needs to be managed with care and consideration: you must be sure that the conditions at the reintroduction site are suitable; you have to know that these conditions, and any necessary management, can be maintained in the long term; and there needs to be potential for the introduced species to spread into a wider landscape. You also need to be able to monitor the reintroduced population to be sure it is increasing and spreading, and to learn lessons that can be applied to other reintroduction projects. Fortunately, Ranscombe Farm Reserve clearly fitted the bill. Interrupted brome was historically found on farmland, and this included arable fields which suited its annual lifestyle. Ranscombe Farm was already being managed as a nature reserve, with a long-term commitment by a conservation organisation in the shape of Plantlife. There was a large area of farmland into which the plants could potentially spread. And staff and volunteers were on hand to monitor the – hopefully – increasing population. Ranscombe Farm was also well within the known historic range of interrupted brome – it had been found growing near Cobham, some 3 miles west of Ranscombe Farm, back in 1912, and the last known Kent site was at Ladd's Farm, 3 miles south-west of Ranscombe Farm, in 1936.

So, In October 2015, Fred came to Ranscombe with about 3,000 seeds of interrupted brome, which were sown into two trial plots: one of around 20m² (about 215 square feet) which received most of the seed, and the other just 7.5m² (about 80 square feet) which got the left-overs. The seed was lightly raked in, and then we crossed our fingers and waited. Fortunately, the seed germinated very readily, the seedlings survived, and by the following June the larger plot was sporting some 9,000 fresh, green flowerheads, while the smaller plot had another 150. Over the next three years, and without any further addition of seed, the population in the small plot increased tenfold, with around 1,500 flowerheads across an area of 44m² (475 square feet) – though, in contrast, the originally much bigger population in the larger plot dwindled to about 1,100. By collecting a proportion of seed each year and sowing it into new patches of ground, the population has been increased further, with more than 12,000 flowerheads produced in 2021. Counting flowerheads is easier than counting individual plants, as the flowering shoots all come up from the base, so where plants are growing closely together, it can be difficult to tell if you have one plant with three flowerheads or three plants each with one flowerhead. But 12,000 flowerheads is likely to represent at least 6,000 to 10,000 plants.

Ranscombe Farm Reserve is one of five reintroduction sites, so it looks as if interrupted brome does have a future on nature reserves, or on private land where management can be targeted to its needs: one of the longest-established reintroduction sites is private farmland. But is it going to move out and become re-established more widely in the wild – and was it ever really a plant of wild landscapes in the first place? The experience at Ranscombe Farm suggests that interrupted brome is not a plant which wanders about much – patches of plants have spread only slowly, and even then much of the movement has been down to machine cultivation (sometimes a hand-steered power rotavator, sometimes a tractor-mounted harrow) which dragged soil and seeds along with it. There's been no evidence so far of seed making any substantial leaps and setting up new populations away from the originally sown sites.

Historically, interrupted brome was only ever found in association with farmed land – and then mainly in fields sown with sainfoin *Onobrychis viciifolia* to provide a hay crop. Sainfoin is a handsome, pink-flowered plant in the pea family, and like its relatives, it is able to fix nitrogen from the atmosphere and use this to make proteins. As a result, sainfoin provides good-quality fodder for livestock, and it has been grown for animal feed for centuries: its name is actually French for 'healthy hay'. But it is not a crop that keeps going for very long, and growing sainfoin as a crop involves resowing it every few years to ensure it remains vigorous. It is assumed that interrupted brome originally got from field to field and farm to farm by hitching a ride amongst the sainfoin seed and being accidentally sown alongside it: one important scientific paper reports interrupted brome seed being found amongst sainfoin seed when it was tested for contaminants.

Which is all well and good, except that interrupted brome is considered to be an annual, able to germinate in either autumn or spring, and then flowering and setting seed in the following summer. There is also an assumption – borne out by experience – that some form of light cultivation is necessary to provide the bare ground needed for the seeds to germinate and to reduce the competition from other plants. In contrast, sainfoin is a perennial, being sown in late spring or early summer and not cropped till the second year; after this there would be no further cultivation and the sainfoin would be cut for hay in midsummer for three or four more years before the crop declined and it had to be resown. The two systems hardly seem to match: interrupted brome and sainfoin seem unlikely partners. However, there have been some observations made at Ranscombe Farm Reserve, and on interrupted brome plants in cultivation elsewhere, which may shed a little light.

In 2019, we did some work at Ranscombe Farm Reserve to see whether interrupted brome would grow as well on clay soils as on chalky soils. Seed was sown in the spring of that year in the margin of a field on clay-with-flints, but the experiment was abandoned when it seemed that the seed sown on the clay soils had failed – all we had was a dense growth of various grasses and no sign of interrupted brome flowers. The field margin was cut in late summer, as part of routine management, and the test plots were forgotten. By chance, the margin did not get cultivated over the following winter and spring, as it normally would have been, and in May we were surprised to find the abandoned test plots were chock-a-block with flowering interrupted brome plants. They were big plants, too, and with many more flowerheads on each plant than we were used to seeing on the plants which grew from autumn-sown seeds – perhaps not surprisingly, as autumn-sown seeds germinate in around October and then flower the following May, after just seven months of growth, and that in the coldest part of the year. The spring-sown plants may have had twelve or thirteen months of growth between germination and flowering, including a whole summer.

I know from growing interrupted brome at home that spring-sown plants can get pretty chunky – maybe 20cm (8 inches) across and many times the size of the spindly plants that arise from autumn germination. I've also seen how interrupted brome can sow itself into a short-mown lawn and survive repeated mowing only to flower if it is allowed a few weeks without a cut. So perhaps it is not such a strict annual after all, but able to flex its lifestyle to suit circumstances. Perhaps what happened with sainfoin crops was something like the following. Interrupted brome seeds get sown accidentally alongside sainfoin seeds in spring, and quickly germinate. The plants grow, but occasional cutting during the first summer to control weeds (as is the recommendation for establishing new sainfoin crops) means they do not get to flower until the following May. The sainfoin is then cut in June for hay, leaving short, open vegetation where the next generation of interrupted brome can get going. Although this is just guesswork and hasn't been tested, it might explain the note in a 2002 paper in *Watsonia* (the scientific journal of the BSBI) about a farmer in the mid-20th Century complaining about interrupted brome that "he cannot get rid of the weeds. It doubtless occurs each year in his sainfoin". It may also explain why, as reported in the same paper, the eminent 19th Century and early 20th Century botanist, George Claridge Druce, regarded interrupted brome as a biennial or "sub-perennial".

Whatever strategies it employed to survive, it seems that interrupted brome used to do very well in many places where it occurred, with the *Watsonia* paper stating that it "often appeared in abundance and was relatively persistent in its localities". Its sudden decline in the early to mid-20th Century was no doubt driven by improved cleaning of agricultural seed and by sainfoin falling out of fashion as a fodder crop. But what explains its sudden appearance? After all, in less than a century from its discovery in the 1840s, interrupted brome had been recorded in more than twenty counties in Southern and Eastern England. Some authors have argued that this rapid spread is good evidence that the plant was an accidental introduction from abroad, as newly arrived species – if given a means of moving around – will often spread rapidly as they find and take advantage of suitable habitats and locations. But if interrupted brome does hail from somewhere outside the British Isles, then no-one has yet discovered where this original population might have been.

The alternative explanation is that interrupted brome didn't actually exist at all until the last couple of hundred years or so, but arose as a completely new species through a chance mutation or variation of some closely-related grass, and then found that the farming practices of the time suited it very well, thank you. The 19th Century was a time of massive urban expansion in Britain, and the towns and cities ran on horse-power: horse-drawn carts and drays, horse-drawn omnibuses, horse-drawn cabs. The growth of great new urban areas created a huge demand for hay, until the rise of the internal combustion engine rapidly drove horses out of business. So perhaps this was a case of a spanking-new species arising and finding it could take advantage of a growing industry of sainfoin-growing, only to lose its support system as that industry declined and died.

This is not an outrageous idea. Something similar has been suggested for other species of brome, which may themselves have evolved as farming spread through Europe over the last 10,000 years. It has even been suggested that the common poppy, our most familiar cornfield weed, was not originally a wild plant at all. Instead, it may have actually evolved in the very earliest farmed fields of the Fertile Crescent as a response to farming itself. The common poppy may have started out as a genetic variation of one of a group of closely-related, native, wild poppies of the Eastern Mediterranean, or possibly as a hybrid between two or more of these species where their ranges overlapped (and still overlap) somewhere around modern-day Israel or Palestine. Finding itself particularly suited to the emerging culture of arable farming, this new type of poppy then flourished and increased and spread in a way that its wild precursors just could not manage. Eventually it settled down in its new home and stabilised as a separate, and hugely successful species: while its close relatives and likely ancestors remain in Turkey, Israel, Palestine and Egypt, common poppy has accompanied its human companions to every continent except Antarctica. This influence of human activity on the evolution of plants is so widespread that a term has been coined: "anecophytes", meaning plants which first arose under the influence of human activities, and which do not have a home in a truly natural environment.

What does that mean for conservation? Because I don't know about anyone else out there, but thinking about all this makes my brain hurt. If a whole new species appears because of the ways in which humans have

Sainfoin - once grown as a fodder crop for feeding livestock. Cultivation of sainfoin seems to have provided a perfect habitat for interrupted brome.

changed the environment, then does it matter if we change things back and that species disappears? Wouldn't such a loss actually represent a bit of a clean-up, removing some of the human impact on the wild world? Interrupted brome may have appeared because people in Britain needed lots of hay to support urban living, and chose to grow sainfoin as a hay crop; then it disappeared because we stopped growing so much hay and ditched sainfoin. Should we care about this? After all, if interrupted brome really did come and go as a new species in such a short period, then how many other new and distinct species may have come and gone, unseen and undescribed, since people first began altering the environment all around the world?

What about the species that have come and gone as a result of wholly natural changes to the environment? As I have talked about above, evolution isn't something that happened in the past. Plants are changing, hybridising and evolving all the time, so that even the idea of a species is blurry at best, with plants like eyebrights being prime examples of this. This is fascinating and exciting, but if evolution is continuing, then extinction must be continuing as well. There is a very strong argument that nature conservation should allow evolution to continue – the natural environment is, after all, not a static museum – but does that mean that we should always be relaxed about extinction if it happens naturally? And, given the extent and intensity of human influence on the planet, could we ever be sure that any extinction was entirely natural?

These are big questions which possibly – even probably – have no clear answer. But they are not distant issues that happen in faraway, wild places. What Ranscombe Farm Reserve shows is that by looking at the plants in just one 600-acre nature reserve – or on our doorsteps, or under our noses – we can see the processes that drive evolution and the results of those processes, and that even a single site like this can throw up huge issues about what we want or need to conserve, and why.

Chapter 9 - Spineless characters

Just as the word 'meadow' in 'wildflower meadow' can mean different things to different people, so can the 'wild' bit. We all have our own interests, preoccupations and ways of thinking about the world, which means that we may have greatly varying definitions of words like wildflower, rewilding, even wildlife. It is still easy to buy, from UK firms, 'wildflower' seed mixes that include plants like Californian poppy *Eschscholzia californica*, pot marigold *Calendula officinalis* or love-in-a-mist *Nigella damascena*, all of which may sometimes be found in the wild in Britain, but none of which would meet my definition of 'wild' because in my experience they don't seem to establish self-sustaining populations away from gardens. In the same sort of way, I have always had my own idea of what constitutes 'wildlife': to me, this means all life – other than humans – that lives in the wild. In other words, wildlife would be all non-domesticated plants and animals and fungi and bacteria and all the other living things that do not fit any of those categories. But a lot of people think of 'wildlife' as just referring to animals. It turns out that even the dictionaries cannot agree on this: some, like the online Cambridge English Dictionary, include plants as part of wildlife, while for others, such as the Concise Oxford on my bookshelf, only animals qualify.

Notice, though, that when the definition of wildlife is narrowed down, it is the plants that get missed out. No-one is talking about wildlife and meaning just wild plants. It is as if plants are a second-class type of life or just not viewed as wild. This tendency to see plants as less important or even to ignore them completely in favour of a focus on animals has been called 'plant blindness'. In the past, this is something that I would have put down to television's tendency to limit its coverage of plants to gardening programmes – where the plants are strictly in captivity – while wildlife programmes are almost always all about the animals. But then, when you've got a mechanism for broadcasting moving pictures, it is understandable that you want your subject to move around a bit. Of course, plants do move, and not just through the process of growing bigger, nor only in the showy ways in which a Venus fly-trap *Dionaea muscipula* snaps shut. All sorts of plants are able to twist and bend and stretch: take, for instance, the ways in which the shoots of some climbing plants wave around slowly to find something solid to climb up. For instance, a tendril of white bryony *Bryonia dioica* (a wild member of the cucumber family) will stretch and turn in the air as it seeks out a support to cling to: once it touches a twig or plant stem, the tip of the tendril grabs hold of it. Then the whole tendril coils up like a spring to pull the plant in nice and tight. If you look at a growing white bryony plant, you can see individual tendrils at all the different stages of this process: you just can't really watch it happening in real time – it takes hours, and would be taking the idea of slow-TV to a new extreme.

However, it turns out that television is not the baleful influence that I believed it to be. Some scientists actually regard plant blindness as being part of the way the brain works: because our eyes take in so much visual information, the brain has to have a way of quickly spotting what is most important amongst all those images. So we are programmed to pay attention to movement, to things that clearly stand out from the background, and to potential threats. Which means that we automatically pay more attention to animals – it is built into the way we work.

Understanding how our brains see the natural world is important and useful, but plant blindness does still remain prevalent in a lot of conservation thinking. Plants tend to get defined a lot by their usefulness to animal species. Fumitories, as an example, have recently got a lot of attention because their seeds have been found to be a favourite food of the rapidly vanishing turtle dove *Streptopelia turtur*, and it is likely that more people are aware of this than would know that some fumitory species are rare and threatened in their own right. Or there's the low-growing horseshoe vetch of chalk grassland, which is seen as important because it is the food-plant of the caterpillars of the chalkhill and adonis blue butterflies. Which it is, and there is nothing wrong in understanding this relationship. But no-one thinks they have to justify the value of the butterflies in terms of the other species they support (the fly *Aplomya confinis*, anyone? Its

Two downland butterflies: chalkhill blue (left) and adonis blue (right).

larvae feed on the caterpillars of chalkhill and adonis blue, as well as other blue butterflies).

Possibly the biggest conservation issue where there is a mismatch between plants and the animals they support is the well-publicised decline in pollinating insects. This is an important issue and a big one. It is also complicated and has become badly tangled up in some of the popular media. Now, I am not as qualified as many other people to unpick the various factors affecting pollinating insects and their relationships to plants, but I have spent a fair bit of time looking at bees, so please bear with me while I have a go. The media focus on pollinators seemed (to me, at least) to start with concerns about declines in numbers of many bee species, especially some species of bumblebee, which then got mixed up with separate stories about threats facing honey bees *Apis mellifera*. Now, some of the things that affect honey bees also affect bumblebees (as well as the rest of the approximately 275 species of bees in the British Isles) such as the widespread use of certain insecticides and some shared diseases. However, other problems, such as the *Varroa* mite, are specific to honey bees. In any case, honey bees are mainly a domestic animal in Britain, kept and managed for production of honey. Although honey bees can and do live in the wild, they are only wild animals in the same way as, for example, the feral goats that live and breed in Snowdonia, and which also originated from escaped domestic animals. As well as honey-production, honey bees are important for pollinating crops, so their health and well-being is as important as the health and well-being of other domestic livestock, such as cattle or sheep. But the problems affecting honey bees are not direct nature conservation problems, any more than is foot-and-mouth disease in cattle.

Then there is the question of why pollinating insects are important. It is certainly the case that a lot of the fruit and veg we eat is dependent on pollination by honey bees and wild species of bee (in fact, it may be the species other than honey bees that do the bulk of this work). Pollination is also important for wild plants: many wild plant species have gone to a lot of effort to attract insects and to get them to move pollen from one flower to another. Of course, the insects do not know that they are doing this – they are visiting flowers to feed on energy-rich nectar, or to eat or collect protein-packed pollen (or, like some orchid pollinators, they are just being tricked into visiting flowers). But there is an interdependence here, though, on balance, the pollinators probably need the plants more than the other way around. After all, many plants are wind-pollinated and lots are able to self-pollinate, but bees are totally reliant on plant pollen for their survival: it is what their young are reared on, so no flowers means no bees at all.

On top of this, there are a lot more insects than just bees involved in pollination. There are wasps, flies, beetles, butterflies, moths and perhaps a few other groups which feed on nectar or pollen or both at some stage in their lives. As a result, a lot of the work that is done in the name of pollinator conservation only manages to address part of the problem. For instance, one of the options that farmers used to be able to take up through previous agri-environment payment schemes (schemes such as Higher Level Stewardship) was to plant pollen and nectar mixes along the margins of fields: these were seed-mixes which included plants known to provide food for butterflies and bumblebees. And these certainly work as far as they go. But the number of species they support is limited: plants that are good for bumblebees are not necessarily much use to other bee species. In fact, a study published in 2022 showed that three species which are often targeted for removal or control on farmland – common ragwort *Jacobaea vulgaris*, creeping thistle *Cirsium arvense* and spear thistle *Cirsium vulgare* – supported twice as many species of bees, and twice as many bees in total, when compared with the plants which farmers are asked to plant for pollinators. I suspect that something similar is likely to be true of the mixtures of colourful, annual flowers planted along roadsides or in parks to help 'save the bees', which often include a mix of different flowers from around the world: these sorts of plantings will attract insects and may supply lots of nectar but they cannot help those bee species that are fussy about where their pollen comes from.

Do not get me wrong, though: I can see the value in these sorts of actions. If one species or a small group of species is declining rapidly, as bumblebees are, then doing something very specific for them – such as planting the kinds of flowers they need – is justified. If it is an emergency, then emergency action is needed. And you can argue that sowing a mix of nectar-producing garden plants to help pollinators in a park is not as good as creating more natural flower-rich grassland, but it is a lot better for wildlife (and nicer to look at) than just a patch of short-mown grass. But in the long-term, the only way to protect the future of the widest possible range of pollinating insects – whether as a way of helping pollinate crops or wild plants, or, just as importantly, because these insects are important and amazing in their own right – must be by protecting, restoring and recreating real wildflower habitats. What we need is grassland, heathland, moorland, marshes, coasts, cliff-tops, woodlands and towns to all be naturally full of flowers.

On this basis, you would hope that a site like Ranscombe Farm Reserve would be good for pollinators, otherwise the case I have just argued would be completely undermined. Ranscombe Farm's management is focused on the conservation of wild plants, and the site is big enough to hold a range of habitats including arable, grassland and woodland. Plantlife has used the slogan "The British Countryside: Save it with Flowers", on the assumption that, as plants are at the base of every food-chain, what is good for plants will also be good for all the other wildlife that depends on them. At Ranscombe Farm, there has been no deliberate planting for pollinators, and, with the exception of putting out some bee hotels and creating a couple of small scrapes, there has been no work at all targeted at encouraging bees and wasps. So the site could be seen as something of a test of the Plantlife approach.

Fortunately, it turns out that Ranscombe Farm Reserve really is pretty good for bees and wasps. To some extent, this may be helped by the site's location close to the 'East Thames Corridor' – this being the name given to the long stretches of the north and south banks of the Thames east of London, which are known to support a rich mix of insects and other invertebrates. The East Thames Corridor is important for insects for a number of reasons: lots of 'brownfield' post-industrial sites; plenty of dry, open habitats on sandy and gravelly soils; the presence of flower-rich sea-walls and patches of marshland; and higher-than-average sunshine totals. Ranscombe Farm sits a short way outside what is strictly regarded as the East Thames Corridor, but does have some species in common with it. These include, for example, the rare five-banded weevil-wasp *Cerceris quinquefasciata*, long seen as one of the important species of the corridor's sandy soils but at Ranscombe Farm recorded from both chalky arable field margins and chalk grassland.

Nonetheless, the five-banded weevil-wasp (more of a description than a name – it has five yellow bands on its abdomen and stocks its nest burrows with small weevils to feed its young) is just one of more than forty wasp species which we found during survey work at Ranscombe Farm Reserve between 2012 and 2019. Ninety species of bees (around one-third of all the bee species in Britain and Ireland) were also found in this period, which clearly reflects the wide range of flowers on the site, particularly as some of these bees were fussy species which only collect pollen from certain species of plants. For example, they included the gold-tailed melitta *Melitta haemorrhoidalis* and the tiny, black small scissor bees *Chelostoma campanularum*, both of which visit bellflowers for pollen – they are known to visit all three of the bellflowers which occur at Ranscombe Farm: nettled-leaved bellflower *Campanula trachelium*, clustered bellflower and harebell.

Fussier still are those bees that only collect pollen from a single species of wildflower. The red bartsia bee *Melitta tricincta*, as its name suggests, only collects pollen from the pink-flowered, semi-parasitic plant red bartsia. This bee is largely confined to the chalk of Southern England but is relatively widespread there – the males can be easy to spot as they patrol through patches of red bartsia on the look-out for females. Much less common is the large scabious mining bee *Andrena hattorfiana*, an

The large scabious mining bee - a scarce species that only collects pollen from scabious flowers - in this case, field scabious growing by the Ranscombe Farm Reserve car park. Photograph: R I Moyse/Plantlife.

impressively big, dark bee which collects pollen from field scabious *Knautia arvensis* (though it will also visit small scabious): the pollen from these blue-flowered plants is actually pink in colour, so with a good load of field scabious pollen packed into the pollen baskets on its hind legs, the bee can look like it is wearing a pair of pink pantaloons. Ranscombe Farm also provided the first Kent record of the bryony mining bee *Andrena florea*. These mid-sized mining bees collect pollen from the pale-green, five-petalled flowers of white bryony, and were until recently very localised in their UK distribution. Their presence at Ranscombe Farm Reserve seems to have been due to a sudden expansion of the population on the Essex side of the East Thames Corridor, and the species has since spread across much of North and East Kent.

One of the most interesting discoveries was just how good Ranscombe Farm's arable fields were for bees and wasps. I mentioned above how farmers are encouraged and supported to deliberately plant field margins with flowers which provide nectar and pollen, but this has never been done at Ranscombe Farm Reserve, and many of the arable field margins are managed for plants simply by cultivating them and allowing natural regeneration of the annual cornfield weeds. Some of these margins are alongside trees and scrub, some next to footpaths, and others alongside grassy strips, but nearly 80 species of bees and wasps were found on and around these margins, which strongly suggests to me that there is a case to be made for less reliance on sown pollinator strips and more emphasis on natural regeneration of vegetation. If nothing else, there should be more studies of the benefits to pollinators of letting wild plants take the lead on arable land, and not simply opening a seed-packet – at least not as the first step. This idea is backed up by a scientific paper by researchers in Germany who found a particularly high number of bee species in arable fields which had been left unmanaged for two years so that the vegetation had developed a mix of annual and perennial plants. The paper notes two important factors: first that annual wild plants were far more attractive to bees than they had expected, and that the greater the number of flower species present, the greater the variety of bee species. This relationship between plant diversity and bee diversity was also suggested by some data collected at Ranscombe Farm by the Wildlife Study Group volunteers: we found that the field margins with higher numbers of bee species tended to be those with a greater variety of wild plants, although our work was not conclusive.

However, bees and wasps need more than just flower-rich habitats: they also need the right sort of habitat structure. Sunshine and shelter from strong winds are important, but more than anything, they need somewhere to nest and reproduce. To understand this, it is important to know something about the lifestyles of these insects, and this is where honey bees are always a terrible example as they do lots of things that other bees do not. For example, the two big things we all know about honey bees is that they are social insects that live in colonies, and that they make honey: both of these are unusual behaviours in bees. The story of a honey bee colony is that it consist of a queen – the only member of the colony to reproduce – supported by a very large number of closely-related workers, and it is the workers that look after the colony and rear the young from the eggs laid by the queen. All the workers are female, and when males are produced, their only job is to mate with new, young queens so that they can establish new colonies. Colonies are also perennial: all that stored honey is there to feed the queen and workers through the winter when it is too cold to go out and flowers are in short supply.

Only a relatively small number of the other bee species in Britain are social. Mostly they are what is known as 'solitary' bees: their nests have no worker bees and are just made by individual females working on their own. The nests are normally in hollows and cavities: burrows dug into the ground, or holes in wood or sometimes in masonry. Each female, working alone, will lay eggs in the nest – each egg in its own cell – and provide the egg with pollen and a little nectar which will feed the larva after it hatches and as it grows to maturity. Job done, the female will then die. There may be one or two generations in a single year, but only the young bees in the nest will live over the winter to emerge the following year. With no colony to feed over the winter, there is no need for honey to be made and stored.

A few species of mining bees that dig nest burrows have evolved to be social and do have small numbers of workers, and all the bumblebees are social insects with queens supported by a big team of workers. But even then, their colonies do not last over the winter. In the bumblebees, new queens, having emerged and then mated in the previous summer, hibernate through the winter while the rest of the colony – the old queen, all the workers and any males – die before winter sets in. It is true that with increasing temperatures, and especially in towns, some bumblebee species can avoid hibernation and simply keep producing new colonies throughout the year, but each colony is still relatively short-lived and bumblebees only store small amounts of nectar to tide them over occasional periods of poor weather.

Wasps are similar. There are a few species which are social, and these are the ones we are most familiar with: the so-called 'yellow-jackets' that bother us at picnics and build large paper nests in outbuildings and attics, or in holes in the ground, or in trees. Like the bumblebees, these colonies, though they may be large, die off completely at the end of summer, with only the new queens surviving in hibernation over the winter. But most wasps are solitary, like most bees, and also nest in burrows or hollows in walls or timber. Their relationship to wild plants is different, though, as they are carnivores: while bees feed their growing larvae on high-protein nectar, wasps stock nests with animal protein, often in the form of other insects (though a worker of one of the social wasp species did once cut and carry off a tiny piece of stilton cheese from the ploughman's lunch I was eating outside my local pub. And, yes, I sat and watched this happen). Adult wasps, though, are as energy-hungry as bees, and visit flowers to drink nectar, which is when they end up playing an important part in pollination.

This means that good habitat for bees and wasps has to have plenty of suitable nesting places as well as plenty of suitable plants. Good nest sites might be areas of short grass or bare ground where burrowing species can dig, or dead wood with beetle holes, or the broken ends of hollow twigs, or other suitable cracks and crevices. And, because bees and wasps need to keep warm, these potential nest sites must get plenty of sunshine. At Ranscombe Farm, we found that a lot of suitable sites were being provided just as a by-product of our work for plants. I have mentioned in a previous chapter how the open space created and maintained for plants at Mill Hill also provides suitable nesting habitat for the wasp, *Argogorytes mystaceus*, which pollinates fly orchids. In Cobham Woods, the creation of another wide, sunny woodland ride allowed the sun to beat down onto a low, sandy bank, and this rapidly became home to a number of bees and wasps, including the green-eyed flower-bee *Anthophora bimaculata*, a buzzing, darting insect with striking, iridescent eyes and more often associated with heathy or coastal habitats. The delightfully named small shaggy bee *Panurgus calcaratus* started using bare ground around a gateway created when we fenced an area to establish a new, grazed pasture, and also seemed to appreciate a long, bare scrape intended to encourage plants of disturbed ground. But it is often on footpaths that bee nests are most obvious, as a range of species like to tunnel into the compacted soil of a path surface, the nests often being made obvious by the little cone of spoil thrown up around the burrow entrance.

In fact, on some footpaths, you can sometimes see a whole ecosystem in action, based around nests of the sharp-collared furrow bee *Lasioglossum malachurum*. These small, dark bees often form aggregations of burrows in the compacted soil of the footpath, supporting themselves on a wide range of nearby flowers. Patrolling above the burrows may be the black-and-yellow forms of ornate-tailed digger-wasps *Cerceris rybyensis*, another burrowing species but one which catches small bees for its larvae to feed on: their aim is to grab some of the furrow bees as they go back and forth from their own burrows. Also sneaking about can be box-headed blood bees *Sphecodes monilicornis* – these are one of a number of species of cuckoo bees, which creep into the nest of another species and lay their eggs there. If you add to that the plants these species depend on – the sharp-collared furrow bee will visit a wide range, including blackthorn and other shrubs in spring, and the ornate-tailed digger-wasp has been known to take nectar from hogweed *Heracleum sphondylium*, wild carrot, and yarrow *Achillea millefolium* – then you can start to see a whole web of ecological relationships which can only happen where there is a mix of plants growing in a range of different habitats from completely bare ground to grassland and scrub.

Watching bees and wasps going about their lives can be entirely engrossing. It was at Ranscombe Farm that I first saw a red-tailed mason bee *Osmia bicolor* sorting out its nest. These are very smart bees – the females are deep black on the head and thorax but their abdomens are a rich red. They are found on chalk grassland in Southern England, and they nest, perhaps surprisingly, in empty snail shells. A few bee species do this, using the shell as a ready-made hollow in which they form and provision a series of cells before plugging the entrance. Red-tailed mason bees go one step further and hide the completed nest under a cover of bits of leaf, pine needles, or – as in the case of the bee I watched – short lengths of dead grass. The female bee flew back and forth, coming in each time with a single piece of dried grass an inch or so long (and so considerably longer than the bee herself) which she placed carefully over the snail shell before heading off for another bit of her thatch material. It gives an appreciation of the complexity of behaviour that can be coded into the instincts of an animal with such a tiny nervous system. You can be tempted to imagine that the creature has some sort of internal life, perhaps even feeling a sense of satisfaction when it completes its work – and who's to say that its instinctual drives don't work in that way?

It is clear, then, that having a wide diversity of flowering plants and a varied and complex habitat structure are both important for the conservation of bees and wasps. But there is another aspect of the vegetation that matters, and that's having not just diversity but a huge abundance of flowers right through the spring and summer and into early autumn. Different species of bees and wasps are active at different times of year: some are early risers, like the chubby, fuzzy hairy-footed flower bees *Anthophora plumipes* – the females all in black, the males in brown – that appear in spring, while others, like the ivy bee *Colletes hederae* don't appear as adults till late summer. Lots more are variously in between, some active for just a few weeks, others for months. So it makes sense that you can only have more bees if you have lots of flowers available right through the warmer months. But for bumblebees, this is a matter of life or death.

The ivy bee specialises in collecting pollen from ivy flowers.

Bumblebees are social insects. Each year, the queens emerge in spring from their winter hibernation and start founding new colonies. For a while they work on rearing the first workers, before retiring to the nest to lay more eggs while the workers rear the young and generally look after the place. The colony grows in size over a period of weeks, but it is only later in the season – after perhaps two months

The brown-banded carder bee is one of Britain's scarcest bumblebees. Here, it is enjoying nectar from red clover in Brockles Field. Photograph: R I Moyse/Plantlife.

or more – that a new generation of queens and males is produced. If the colony does not get to that stage, the whole enterprise will have failed and there will be no succeeding generation to start the cycle again in the following year. So a colony can only be successful if there is sufficient nectar and pollen available to support a large number of bees for several months – and colonies may consist of anything from a few dozen workers to several hundred, depending on the species. At Ranscombe, we found twelve different bumblebee species, which included some of the cuckoo bumblebees which take over the nests of other species, but which still rely on the army of workers in the invaded nest. Bumblebees would use the wide range of flowery habitats, from woodland rides, to grassland, to flower-rich arable. Most exciting was the regular presence of the brown-banded carder bee, a glowing, ginger teddy bear of a bumblebee that has declined massively in Britain and only hangs on where there are extensive areas of flower-rich grassland.

Bees and wasps are not the only insects that are dependent on the abundance and diversity of plants, and a varied and complex habitat structure. Butterflies and moths all feed on plants when they are caterpillars, with a few exceptions such as those that feed on detritus in birds' nests or chew their way through woollen clothing. And, as adults, they take nectar from flowers, though some will also feed on tree sap or on the honeydew excreted by aphids – and there are even some moths that don't feed at all beyond the caterpillar stage. But, by-and-large, plant and habitat diversity equals diversity in butterflies and moths, and at Ranscombe Farm Reserve more than 330 moth species have been recorded so far, as well as 35 species of butterfly.

For the visitor to Ranscombe with an interest in butterflies, open spaces in woodland – especially in Cobham Woods – are the places to look for silver-washed fritillaries *Argynnis paphia* in summer. Large and stately, and a rich orange-brown with black markings, these butterflies are not especially common in Kent, but they do like sunny, sheltered woodland rides where there are plenty of the common dog-violets *Viola riviniana* their caterpillars feed on. Also enjoying the warmth and shelter of the rides are the smart white admirals *Limenitis camilla* – black with broad, white bars – though these lay their eggs on honeysuckle in nearby shady woodland. Moths can be harder to spot during the day, though dingy footman *Eilema griseola* – pale grey and shaped a bit like a pumpkin-seed – can be found feeding on ragwort flowers.

At Ranscombe Farm, the grassland probably has the most interest for the butterfly lover. In early summer, the rather plainly coloured dingy skipper and the smartly spotted grizzled skipper may be found in the sunny grassland at Brockles Field or even in the small patch of grassland above Kitchen Field. Both are threatened species (classified as Vulnerable in the British Red List) associated with sunny downland where they can find their favourite caterpillar food-plants – common bird's-foot-trefoil and wild strawberry, respectively. Plantlife's work to restore Brockles Field as flower-rich chalk grassland seems to have worked especially well for these two species, but the tiny patch of chalk grassland at Mill Hill has its own surprises, with both chalkhill blue and adonis blue butterflies being regularly present in recent years. Grassland is also a good place to look for day-flying moths, of which there are a good number of species, from the vivid, blue and scarlet six-spot burnet *Zygaena filipendulae* to the somewhat drab burnet companion *Euclidia glyphica* and the Mother Shipton *Callistege mi*. A personal favourite is the slender and scarce moth *Oncocera semirubella*, something of a chalk downland specialist and sometimes given the English name rosy-striped knot-horn – though I prefer the name a friend of mine gave it – rhubarb-and-custard – which suits its mix of yellow and pink colouration. You may disturb these small moths as you walk through the grassland at Mill Hill, for although their foodplant – common bird's-foot-trefoil – is widespread, the little rhubarb-and-custard moth seems mainly to be found in the better bits of chalk grassland.

Even arable has its share of butterflies and moths, with some species specialising on annuals, such as the chamomile shark moth *Cucullia chamomillae*, whose caterpillars eat scentless mayweed and stinking chamomile. But it is the edges that seem most interesting – those places where one habitat blurs into another. Part of the management approach at Ranscombe Farm Reserve has been to deliberately try and create habitat continuity for wild plants by, where possible, eliminating sharp boundaries between one type of habitat and another. As a result, grassland butterflies can move easily up into the wide, sunny woodland rides or into flower-rich arable. One of the best butterfly experiences I've had at Ranscombe Farm was watching the courtship flight of silver-washed fritillaries – the male looping-the-loop around a fast-flying female – as they came out of Cobham Woods and headed off across the flowery arable land of Kitchen Field. It is an amazing piece of behaviour to watch: the female flies in a straight line as the male swoops beneath her, coming up in front and slowing so that she passes beneath him, repeating this move over and over. But what made it even more rewarding in this case was knowing that the butterflies were treating the woodland, grassland and arable not as separate habitats but one continuous piece of great butterfly landscape.

Butterflies and moths also connect Ranscombe Farm Reserve into a very much wider landscape. Ranscombe Farm's dingy and grizzled skippers and silver-washed fritillaries must be part of a larger population using the surrounding parts of the North Downs, or they would have been unlikely to find their way to the reserve in the first place. This is even more true of the chalkhill blue and adonis blue butterflies using the little patch of grassland at Mill Hill, as a small patch of habitat like this can only ever support a tiny butterfly population, making it always vulnerable to local extinction. So it seems more than likely that these two stunning species – chalkhill blue males are a lovely, pale, powder-blue, while male adonis blues are a dazzling, vivid blue unlike any other British butterfly – are at Ranscombe Farm as part of wider, more loosely-connected populations on other flower-rich downland within a few miles' radius. That is unless they were deliberately released here from individuals reared in captivity.

Unregulated and unauthorised release of wild animals is a genuine problem for conservation, and is particularly common with butterflies. It is easy to buy caterpillars of lots of butterfly species online, and people regularly release adults into the wild, without letting anyone know they have done so. This is something that might seem like it is helping conservation, but in reality it is a hindrance as it muddies the picture of the real distribution and behaviour of butterfly species. In the case of the adonis blue at Ranscombe, we are dealing with a fairly rare butterfly where its conservation depends on knowing how far individuals are likely to travel to find and colonise new sites: having an understanding of how far an animal travels is essential when planning how to restore and recreate networks of suitable habitat. Ranscombe Farm is five miles from the nearest population of adonis blues at Kent Wildlife Trust's Darland Banks Nature Reserve in the Medway Towns, and nine miles from the next nearest at Queendown Warren Nature Reserve, a site managed by Kent Wildlife Trust and partly owned by Plantlife. So if adonis blue butterflies are regularly seen at Ranscombe, as they are, this suggests they routinely travel at least five to nine miles from existing populations – and in sufficient numbers that some are able to find their way to the tiny bit of grassland at Mill Hill. If it turns out the adonis blues which we see at Ranscombe Farm were just released there without any authorisation, then we will have been led completely up the garden path. Worse still, by fooling us into thinking that adonis blues can travel much further than they do in reality, it may well reduce the incentive to restore the kind of habitat networks that they are likely to need to survive and spread in the long-term.

Dragonflies do not have these complications as people are unlikely to deliberately rear and release them, and so they show very clearly how Ranscombe Farm Reserve cannot be seen as a site in isolation. Perhaps the most interesting thing about the dragonflies (and their close relatives, the damselflies) at Ranscombe Farm is that the reserve has no open water. There is a pond in the garden of one of the private houses in the middle of the reserve, and presumably some ponds in nearby residential areas, but most of the dragonflies and damselflies at Ranscombe must have grown up and emerged as adults well outside the reserve's boundaries.

Yet the role played by Ranscombe Farm in the lives of these insects is an important one. Dragonflies and damselflies are all predatory insects: the young (known as nymphs) live in still or flowing water (depending on the species), where they are voracious predators of other aquatic invertebrates or sometimes even tadpoles or small fish, and as adults, they hunt and catch other insects on the wing. When, sometimes after years as a nymph, an adult dragonfly or damselfly first emerges into the air, it needs to spend the first part of its short life away from water while it becomes fully sexually mature. Newly-emerged females want to avoid male attention while they mature, and males want to avoid competition from other males – which are often aggressively territorial – until they are tough enough to fight back. This maturing period can last from around a week to maybe a month, depending on the species and on habitat conditions, during which time the

A hairy dragonfly taking a rest on the edge of Merrals Shaw at Ranscombe. Photograph: R I Moyse/Plantlife.

best strategy for both sexes is to stay well away from any breeding habitat and to find somewhere with lots of available food to fuel their further development. Ranscombe Farm Reserve attracts large numbers of dragonflies to come and feed. They are seen very frequently in the woodland rides and glades, where the sheltered conditions mean there is plenty of their small insect prey, but also over grassland and even over arable where it is managed for wildflowers. This is a very good indication of how annual cornfield weeds can support small insects which are the basis of a bigger food chain, and I have watched a hobby *Falco subbuteo* – that slender, elegant falcon – flying low over an arable field as it hunted dragonflies which were themselves feeding on other, smaller insects: so a predator feeding on predators feeding on small insects feeding on wildflowers.

Probably the commonest dragonfly species seen at Ranscombe Farm are the common darter *Sympetrum striolatum*, a species with orangey-red males and straw-coloured females, and the blue-coloured migrant hawker *Aeshna mixta*. But the large, blue and green emperor dragonfly *Anax imperator* is also regularly seen, and the aptly-named broad-bodied chaser *Libellula depressa* and hairy dragonfly *Brachytron pratense* have also been recorded. The range of species and sheer numbers suggests that Ranscombe Farm is drawing in these insects from a wide area at least a mile or two in radius, with perhaps most coming from the pools and dykes of the nearby grazing marshes along the River Medway: the relatively uncommon hairy dragonfly, for example, is particularly associated with the marsh dykes that fringe many of Kent's rivers. This provides an important lesson for how we view the natural landscape: Ranscombe Farm may be a dry site, but it still provides feeding habitat for large numbers of insects that will return to wetlands to breed, and whose young are some of the most important predators in freshwater habitats. And, of course, this applies the other way around: the wetlands of the Medway marshes have a strong influence on the ecology of Ranscombe Farm, particularly when the woodland rides and edges are flooded with dragonflies in summer, all feeding on the insects there.

No nature reserve is ever an island, and the connections to other parts of the landscape spread in all directions. Dragonflies and damselflies may link Ranscombe Farm to the Medway marshes, and chalkhill blue butterflies make connections to the chalk slopes of the North Downs, but Ranscombe Farm, as rural as it appears, seems as much linked to the brownfield sites of the East Thames Corridor as anywhere else. I've already mentioned the five-banded weevil-wasp: small, black and yellow, and easily overlooked, but long seen as one of the characteristically important and threatened insects of the Thames margins. But the brown-banded carder bee is also very much associated with the margins of the Thames, and can be found from the outer estuary right up into the built-up parts of London and its satellite towns, where it has something of a local heartland on urban brownfield sites like that at Swanscombe Peninsula. The phoenix fly *Dorycera graminum* – tiny, at just 7mm (⅜ inch) long and Red-listed as Vulnerable in Britain – is also found at Ranscombe Farm. This small but significant insect got its English name from its strong association with brownfield sites, as these old, post-industrial plots of land are often dubbed 'phoenix sites': new wild places that have risen from the ashes and rubble of vanished human activity. What all these different insects suggest is that the distinctions which we make between different types of habitat mean little for the species we want to protect and conserve. If a largely woodland and farmland site like Ranscombe Farm Reserve shares important species with wetland, downland and brownfield sites in the surrounding landscape, what does this say about the ways in which we plan for, identify, protect and manage land for wildlife?

There is potentially a lot more to say about the insects of Ranscombe Farm Reserve: I have just touched on a few of their stories here. But then there is still a lot more to find out about the reserve's insects: those found so far – at just a little over 1,100 species – are surely just a fraction of the total present. Even with those we know, if I had more space and a good deal more entomological expertise, there could perhaps be a lot more to be explored about the ways in which insect diversity and abundance is influenced by the plants present and by the structure and management of the habitat.

What could we learn from flies, for example? So far, something over 280 species have been found at Ranscombe Farm, which must only be scratching the surface as there are more than 7,000 species in Britain. Fifty-seven of the fly species so far found at Ranscombe are hoverflies, almost all of which visit flowers for nectar so that they have an important role as pollinators. In some cases, the relationship with plants is even more complex, as with the large hoverfly *Volucella bombylans*, whose larvae live as scavengers within bumblebee nests, or those other hoverflies with larvae that feed on aphids and so are important as natural pest-controllers in crops. There are also those fly species with larvae that live inside the stems, leaves, flowers or other parts of plants, such as the various picture-winged flies which may attack burdocks, knapweeds, creeping thistle or other plants. Not forgetting the tiny gall-midges whose larvae cause swollen and distorted growths on buds or leaves or elsewhere on the plant, these growths then providing the larvae with the necessary food and shelter (and usually being the only sign that the insects are there at all). Many species of picture-winged flies and gall midges are restricted to just one host plant, so are very dependent on plant diversity

Equally, we as yet know little about Ranscombe Farm's beetles, with just over 180 species found to date. These include 18 species which live in dung and play a key role in recycling it back into the soil, and eight species of long-horn beetles, including some uncommon species; long-horn beetles are also important recyclers as most have larvae which live and feed in dead wood. Adult long-horns, though, visit flowers to feed, so that the presence of suitable flowering plants turns out to play an important role in the ecology of dead-wood and its breakdown.

However, there is one story that insects can really tell us, and that is the one about the colossal changes that are happening to the world's climate. Being in many cases highly mobile and with annual life-cycles that allow rapid expansion in numbers – or just as rapid collapse – insects can very quickly spread or disappear in response to climate change in a way that longer-lived plants and animals do not. A tree or another perennial plant may remain present and maybe even apparently be thriving as the climate changes around it, even if that change means that the plant itself is doomed in the long run: as such, it may be running up its own 'extinction debt'. In contrast, insects are cashing-in or paying-up almost immediately, so changes in populations can sound alarms that might not be heard elsewhere for years to come.

Ranscombe Farm is already a pretty hot place in summer. Whether or not the Medway Gap shares something of the high sunshine totals on the adjacent East Thames Corridor (and which have led to record temperatures in the past), parts of the reserve are sheltered, south-facing and sunny, and can get very hot in summer. This particular microclimate may, for instance, explain the presence of the large, metallic-green beetle known as Spanish fly *Lytta vesicatoria*. This impressive insect seems to have been present at Ranscombe Farm Reserve for some years in small numbers, despite it being more normally a species of Southern Europe, and very rare in Britain. This is, incidentally, the same Spanish fly that has long been considered to have aphrodisiac properties, but this use is not to be recommended as it is actually poisonous and has unpleasant effects if consumed.

Since Ranscombe Farm first become a nature reserve in 2005, there has been a steady stream of new insect arrivals, reflecting the way in which a general increase in global temperature has allowed species from further south in Europe to spread north and west into Britain. Two species of bee which first appeared in Britain at the start of the millennium are now common in the reserve, as they are in the rest of the country: the tree bumblebee *Bombus hypnorum*, which is the only bumblebee species in Britain which will commonly nest in tree-holes and bird-boxes; and the ivy bee, a mining bee similar in size to a honey bee, but which almost exclusively visits common ivy flowers for pollen.

Jersey tiger moth at Ranscombe.
Photograph: R I Moyse/Plantlife.

Also now present is the Jersey tiger moth *Euplagia quadripunctaria*, a large and striking species with forewings which are vividly striped in black and white and with hindwings – hidden when the moth is at rest – of bright red. The guide to moths on my bookshelf was published in 2003 and describes the Jersey tiger as being numerous on the south coast of Devon, and possibly established on the Isle of Wight and near Rye. That was the extent of its British distribution at the time. Less than two decades later, you can sometimes find them out in the day at Ranscombe Farm, just sitting on vegetation in the sunshine. Likewise, the southern migrant hawker dragonfly *Aeshna affinis* was just noted as a rare migrant species when Kent Field Club published Dragonflies of Kent in 2009, but now it is established a breeding species in the county, and has apparently been reported from the woodland rides at Ranscombe.

Sometimes, when talking about the way in which climate change is driving new insect arrivals and population expansions, it looks like good news, especially when species that were once rare and localised in Britain suddenly start being seen more widely. But what does not grab your notice quite so much is the way other species are pushed out as climate conditions become increasingly unsuitable for them. Jersey tiger may now be established at Ranscombe Farm, but what has happened to the garden tiger moth *Arctia caja*? When I was a child, growing up just a few hundred yards from Ranscombe Farm, we used to regularly find their dark, densely hairy 'woolly bear' caterpillars in the garden, and would bring them indoors and rear them till they hatched into insects every bit as big and colourful as Jersey tigers, though the pattern on their brown-and-white blotched wings is perhaps more reminiscent of a giraffe than a tiger. Now garden tiger moths are hard to find, numbers having dropped massively over the last four decades, especially in the South-east – apparently a result of the way climate change is altering our winter weather. Changes in weather patterns may be having a similar impact on hoverflies and craneflies which need moist conditions and cannot withstand extended periods of dry summer weather – yet hoverflies are important pollinators and craneflies food for insect-eating birds.

And while most new insect arrivals are benign, and indeed their arrival and establishment in Britain should be seen as positive evidence of their resilience, resourcefulness and adaptability – something that nature really needs in the face of the tremendous challenge of climate change – a few may pose expensive problems. Oak processionary moth *Thaumetopoea processionea* is a medium-sized, rather plain moth which started creeping up into Britain from Central and Southern Europe during the last couple of decades of the 20th Century. It is now well established in and around London, where it is considered a pest species, partly because its caterpillars can strip oak trees bare of leaves (though the trees do recover) but also because the hairs shed by the caterpillar can cause irritation and rashes, and even breathing difficulties for some people. As a result, the Forestry Commission can require land-owners to take action where there is a risk to visitors – for example where the caterpillars are present over footpaths –

and this means sending climbers up into the trees to remove the protective silk nests which the caterpillars build themselves. A couple of years ago, oak processionary moth was found to have arrived at Ranscombe Farm Reserve, having spread out from its established stronghold in London; taking that necessary action to ensure there is no risk to visitors is likely to prove time-consuming and expensive.

Chapter 10 - Putting some backbone into it

I once very nearly put a toad under the harrow. As a figure of speech, the idea of the 'toad under the harrow' as meaning someone undergoing suffering has fallen out of popular use, though we still talk about 'harrowing' experiences. A harrow is a farming implement for breaking up the soil or for turning the rough surface of a ploughed field into something finer, more even and more suitable for sowing with seeds. The harrow itself may consist of a frame bearing a series of pointed tines, or a series of sharp-edged discs, both pulled across the field behind a tractor or, in the case of the power-harrow I was using, a two-wheeled machine which smashes up the ground using a set of rapidly rotating blades. You might wonder why the suffering animal in the saying would be a toad: after, what is an amphibian doing in a ploughed field, particularly a hot, bone-dry one like Kitchen Field? Yet, there it was, just a foot or two ahead of the blades as I harrowed the field's margin – a small, obviously fairly young common toad *Bufo bufo*. There may be no other moral to this story other than that nature will constantly surprise you: Ranscombe Farm Reserve is after all a generally very dry site, with just one pond in the garden of one of the private houses that lie within the site boundary – and that nearly a mile from where I found the toad.

Because the site has no standing water, amphibians are scarce at Ranscombe. Common frogs *Rana temporaria* do regularly spawn in water-filled wheel-ruts in the wetter parts of Cobham Woods, but these ruts tend to dry out before the tadpoles have grown large enough to metamorphose and leave the water. There are also some oldish records of great crested newts *Triturus cristatus* from a hollow in Cobham Woods which used to regularly hold water for long periods in early summer. Now this shallow depression generally remains dry except in the wetter winter weather, possibly the result of our increasing hot and dry climate, perhaps also because of some illegal quad-biking which tore up the bottom of the pond.

Ranscombe Farm Reserve does have reptiles, though these are mainly just two species: slow-worms *Anguis fragilis* and viviparous (or common) lizards *Zootoca vivipara*. Snakes may be present, although I am only aware of one sighting of an adder *Vipera berus*, and the only grass snake *Natrix helvetica* record during my time at the reserve was a dead individual that may have been dropped by a passing buzzard *Buteo buteo*. An interesting future addition to the site list might potentially be wall lizard *Podarcis muralis*, which is native to Southern Europe but present as an introduced species in several parts of Kent, including – according to reports in the local press – gardens just outside the reserve in the village of Cuxton.

It is when you get to the birds that Ranscombe Farm's vertebrates start to give you more detail about the site and its management, and also how the reserve is impacted by wider environmental changes. The increasing and substantial changes to the world's climate are pushing birds around almost as much as insects, and the northward advance of the Mediterranean gull *Ichthyaetus melanocephalus* is an amazing example. Historically, these were birds of South-east Europe, but they had got to Britain and were breeding by the end of the 1960s and started nesting regularly in Kent in the early 1980s. Today, the islands of the Medway Estuary provide safe nesting for one of the country's largest populations, and birds from these colonies come inland to feed on the open fields of the North Downs. Gangs of the birds in flight seem to be constantly calling to each other, and their call is so distinctive – a sort of yelping, laugh-like 'owww!' – that you can call out "Med gull" without even needing to look up. When you do get a good look at them, they are very smart birds indeed, especially in breeding plumage, when they give the impression that someone has looked at a black-headed gull *Chroicocephalus ridibundus* and said, "I think we can do better than that". Where a black-headed gull has a chocolate-brown hood, the Med gull's hood is deep black and covers the whole head, making a striking contrast with the white ring around the eye. Instead of the black-headed gull's rather slender, dark-red bill, the bill of the Med gull is thicker and a bright scarlet. And Med gulls' underwings are a pure, brilliant white that shines against a blue sky, rather than the dull and shadowy look of a black-headed gull's dark underwing feathers. In other words, it is a species that stands out in both voice and appearance, so I am pretty sure that I am correct in my impression that Mediterranean gulls have only become a regular species at Ranscombe Farm Reserve in the last few years. Whether their presence at Ranscombe Farm reflects increased breeding numbers or simply birds exploring more widely, I cannot be sure, but over a period of a few years, I went from hearing the odd single bird passing over, to spotting groups of birds overhead, and then finally seeing small flocks of twenty or so coming down to feed on Ranscombe Farm's arable fields.

By contrast, birds like the willow warbler *Phylloscopus trochilus* that once bred in the local woods, where its descending-scale song was a common feature of the spring, have shifted their whole British distribution to the damper and cooler north and west, and now only pass through Ranscombe Farm Reserve on migration. Climate change is also a likely culprit for the loss of woodcock *Scolopax rusticola* as a local breeding bird. Around 30 years ago, I remember surprising one of these chunky birds in Cobham Woods on a summer's day – I would not have spotted it if it hadn't moved, as the woodcock's mottled, brown plumage provides superb camouflage against the woodland floor. Why it hadn't seen me sooner, I do not know, as the woodcock's eyes are so placed that it famously has 360° vision. In recent years, they are only seen at Ranscombe Farm in autumn and winter when there is a regular influx of birds to Britain from the continent, and this loss is reflected across Kent, with drier summer conditions likely to make it much harder for these long-billed birds to probe the ground and find the earthworms on which they and their chicks feed.

(Though – and as an aside – climate change does not simply mean the South-east is getting hotter and drier; weather patterns are changing as well, so that there is an increased likelihood of extreme wet weather and extreme dry weather. Which may go some way to explain why local botanist Stephen Lemon was able to find the extremely rare clustered earth-moss *Ephemerum cohaerens* in an arable field at Ranscombe Farm in the last few years. This tiny moss has only three or four other sites in Britain, all in the 'draw-down zone' – the wet, muddy conditions created as water-levels drop – on the margins of artificial reservoirs. But this moss can handle alternating waterlogged and dry conditions, something that would not suit a bird needing to feed in damp soil right through the summer.)

Neither has Ranscombe Farm Reserve managed to buck the big and well-publicised losses that have happened in farmland bird populations. In more than nine years of working at the reserve, I saw just one yellowhammer *Emberiza citrinella* and heard just one singing turtle dove, despite there being some apparently suitable habitat available for both these species. Nonetheless the overall farmland bird picture at Ranscombe Farm is fairly positive as a result of the work for rare cornfield weeds, which provides plenty of the weed-seeds and insects for birds to feed on, and because of a switch to spring cultivation in some of the commercially farmed fields. Spring-sown cereal fields are especially good for skylarks *Alauda arvensis*, as the crops mature later and allow the birds to rear more broods of nestlings before the foliage becomes too dense. Survey work by Ranscombe Farm's Wildlife Study Group volunteers shows that skylarks regularly hold nine or ten breeding territories across the reserve, and that these are concentrated in two key habitats: grassland and spring-sown arable.

Whether yellowhammers will return to Ranscombe Farm remains to be seen, especially as they have a reputation as sedentary birds which are not inclined to explore and settle in new areas. But a recent sighting of corn buntings *Emberiza calandra* provides a ray of hope – these rather plain-looking birds with their big seed-eaters' beaks have been declining in Britain for fifty years but have hung on in farmland to the north and east of the reserve. Only two birds were seen at Ranscombe Farm, and then during the winter, but it will undoubtedly be worth listening out for their distinctive song – a rattly noise like keys being shaken – in future summers.

In general, and across Britain, woodland birds haven't suffered as much as farmland species. But there have still been losses. As described above, woodcocks no longer breed at Ranscombe Farm Reserve, and neither do nightjars *Caprimulgus europaeus*, which, like woodcocks, nest on the ground in open, recently coppiced woodland. But the woods in the reserve are still pretty lively, particularly with birds that appreciate scrubby woodland edges and coppice, or the presence of large, old trees. These include common summer visitors such as chiffchaffs *Phylloscopus collybita*, blackcaps *Sylvia atricapilla*, garden warblers *Sylvia borin* and whitethroats *Curruca communis*, together with resident species like treecreepers *Certhia familiaris*, nuthatches *Sitta europaea* and tawny owls *Strix aluco*. Bullfinches *Pyrrhula pyrrhula* seem to be doing pretty well at Ranscombe Farm, despite a general decline in the country as a whole and especially in the South-east, and are often given away by the rather sad-sounding, single-note calls which these shy birds use to keep in contact with each other. Marsh tits *Poecile palustris*, another species in substantial decline in Britain, are regularly seen in winter and have also been spotted in the summer, suggesting they may still breed in the reserve's woodlands. Even the little lesser spotted woodpecker *Dryobates minor*, a species which has seen a major population crash over the last few decades, may still be present and holding a few breeding territories.

In autumn or winter, you may still find the occasional woodcock, one of the many that come to Britain from the continent, though generally only by accidentally startling it and sending it dashing up from its hiding place on the woodland floor. Autumn and winter are also the time when you might find flocks of lesser redpolls *Acanthis cabaret* feeding in the trees, especially birches, and the arable fields are often good in winter for some other bird species. As well as the two winter-visiting thrushes – redwing *Turdus iliacus* and fieldfare *Turdus pilaris* – meadow pipits *Anthus pratensis* are commonly seen in the winter fields, only becoming noticeable when they take to the air and give their high-pitched, three-note call. Skylarks also gather together in flocks – sometimes several dozen birds together – to feed in the stubbles. On the edges of the fields, where the vegetation is taller, stonechats *Saxicola rubicola* can often be seen perched on the tops of plant stems.

❧

We all like to focus on the doom-and-gloom, but bird conservation has a achieved some really spectacular results, and a number of these are clearly visible at Ranscombe Farm Reserve, even without binoculars. Take, for instance, the stock dove *Columba oenas*: a rather pretty, slate-grey pigeon that in some ways resembles a less beefy version of the woodpigeon *Columba palumbus*, though without the distinctive white markings. When I started watching birds back in the 1970s, stock dove populations had been severely knocked back by the treatment of crop seeds with organochlorine insecticides, something which began in the 1950s with the post-war drive to increase food-production. Stock doves started to recover following voluntary restrictions on the use of these pesticides and their eventual complete banning in 1984, and now these attractive birds are widespread and common. You can see stock doves regularly in the arable fields at Ranscombe Farm – sometimes they are even in small flocks outside the breeding season – and hear their 'hoo-woop' song in some of the woodlands.

Perhaps the most well-known victim of the use of organochlorine insecticides is that powerful and handsome falcon, the peregrine *Falco peregrinus*. Peregrines are great hunters of pigeons. So when they caught and ate pigeons which had themselves eaten treated seeds, the pesticides built up in their bodies to a level where it affected the peregrines' breeding success; they stopped breeding in Kent in the early 1960s. This pesticide poisoning had a similar impact on many different birds of prey, which in the mid-20th Century were still recovering from severe persecution at the hands of Georgian and Victorian farmers and gamekeepers. Back when I first got interested in birds, sparrowhawks *Accipter nisus* had only just started to breed again in Kent following their earlier extinction in the county, buzzards were something you saw if you went west of Hampshire, you travelled to the New Forest to look for hobbies, and, if you were lucky you might see a red kite *Milvus milvus* near the little, mid-Wales town of Tregaron. (Going to Tregaron for the first time was something of a pilgrimage for my family, and our luck was in – we did get to see red kites. Now I see them regularly from the back windows of my house on the Kent Downs.)

Now all these superb birds are back, and it has been amazing to witness what can happen if the needs of a species are understood and conservation work is properly targeted. In this case, this has meant a combination of laws to protect individual bird species, action and legislation on pesticides, and responsible approaches to gamekeeping and predator control; in the case of the red kite, the original population had dwindled to such a small size that focussed and carefully planned reintroductions were needed too. Now Ranscombe Farm Reserve has sparrowhawks and kestrels *Falco tinnunculus*, while buzzards started to breed on the reserve within the last ten years and hobbies are spotted most years. The huge increase in numbers of red kites in England, and their tendency to wander widely, means the sight of one of these birds as it drifts lazily overhead is by no means a rare one. At least one red kite has used Ranscombe Farm as an overnight roost: an early morning set-up for a springtime open day at Mill Hill meant a bunch of volunteers and staff were there in time to see the kite lift up out of nearby woodland and gradually gain height before heading off to do whatever it was planning to do for the rest of the day.

Astonishingly, even ravens *Corvus corax* have arrived back at Ranscombe Farm Reserve in the last ten years. Ravens were absent from the whole of Kent for about 165 years after the last known bird was shot in the 1840s, and it wasn't till 2009 that birds were definitely known to be breeding in the county again. Luckily for Ranscombe Farm, ravens like to nest on cliffs and quickly took to the steep-sided chalk quarries that puncture the Kent Downs in the north-west of the county. Birds regularly pass over the reserve on their foraging expeditions, and I have – just the once – spotted a couple of ravens that had come down to the ground in the reserve, which means they can be legitimately added to the list of Ranscombe Farm's birds.

Peregrines have also taken to nesting on the cliff-like faces of local chalk quarries – and they are on the Ranscombe Farm list too, ever since they discovered that the reserve's arable fields attract flocks of pigeons from the adjacent Medway Towns. These feral pigeons are one of the peregrine's favourite prey items – in the wilds of North-west Britain, peregrines will hunt rock doves *Columba livia*, the species from which domestic pigeons were first bred and therefore the ancestors of the pigeons in cities and towns. I'm still astonished that you can now stand at Ranscombe Farm and watch the high-speed attack of a peregrine as it powers in on its prey, when for so many years this was something that most of us could only hope to see on television.

Of course, none of this makes Ranscombe Farm Reserve particularly special – the return of ravens and peregrines, red kites and the other birds of prey is something that has happened right across the country. You are as likely now to see peregrines in cities as in the country, and I once watched one circling high in the sky while I waited for a train at Strood Station, more-or-less in the middle of the Medway Towns. But I still think it is important because it shows how nature can and does bounce back if we give it a chance and that a real wildlife spectacle – the power of a peregrine's dive, the gangly flap of a red kite's massive, six-foot wing-span – can happen where thousands of people are able to see it first hand and for real, and not just where it has been edited together nicely and put on a screen.

The presence of peregrines is yet another reminder of the ecological connections that run through the landscape and townscape. The peregrines that visit Ranscombe Farm Reserve are most likely to be the birds that use

– and presumably nest in – the old chalk quarry just a short distance from the reserve on the other side of the M2 motorway. This quarry has now been developed for housing, though this has not deterred the birds from making use of the steep chalk cliffs. Major roads are no barrier for a peregrine falcon, but another bird that periodically visits Ranscombe Farm gives a good indication how roads and railways may create impassable barriers for even large birds that are strong fliers. Shelducks *Tadorna tadorna* are large and very smartly marked water birds: black heads and necks, bodies that are mostly white but with black and orange markings. They are mainly a species of coastal marshes and wetlands, but they do breed right up through the Medway Gap, and on more than one occasion I have seen pairs of these birds looking for possible nest sites in Ranscombe. They have even come down into the field called the Knock, probably because they had spotted some holes in the ground that might suit them: shelducks commonly nest below ground in rabbit burrows. Fortunately, they chose not to settle there, because shelducks do not hang around the nest once their ducklings have hatched, but set off to walk the whole brood down to the nearest body of water. In the case of Ranscombe Farm, this would be the Medway, and this walk – though of only about half-a-mile – would involve crossing a very busy road which is difficult for even a human pedestrian, and two railway lines.

The word 'iconic' gets thrown around a lot, and seems often to be used to mean 'easily recognisable' when it more strictly refers to an image or an actual thing that has an important meaning – icons originally being portraits of saints or other religious figures. But badgers are iconic in every sense: most people would recognise one, an image of a badger's face is still used as the logo of The Wildlife Trusts – the association of the various local (usually county-level) Wildlife Trust conservation charities in the UK – and they represent a species which, at least when The Wildlife Trusts were forming, was persecuted and scarce in much of Britain. Badgers, like birds of prey, have bounced back in recent decades after centuries of persecution, and there are now plenty of them at Ranscombe Farm Reserve – the broad tunnels they dig to make their underground setts, their communal latrine pits, their worn, regularly-used trails, and even their footprints can easily be spotted all across the reserve. I am aware that not everyone is pleased to see more badgers, and they have been implicated as one possible cause of the rapid decline in hedgehog *Erinaceus europaeus* numbers (though the People's Trust for Endangered Species and the British Hedgehog

Preservation Society regard habitat loss as a more important factor) but it is worth bearing in mind that a recent population estimate suggests that there are only around 485,000 badgers in the whole of England and Wales – around 6% of the number of domestic dogs in the same area.

Badgers are hard to see, especially in a well-visited site like Ranscombe Farm where they are inclined to remain strictly nocturnal to avoid meeting people: in fact the only times I saw badgers in more than nine years working at the reserve were in the middle of the night after public moth and bat evenings. But at least they leave obvious visible signs – other mammals can be much harder to spot. Hazel dormice leave far fewer signs and are also highly nocturnal, but we know that there are hazel dormice at Ranscombe Farm Reserve, particularly as some areas are regularly monitored by committed local naturalists. We even once caught one on video – entirely by chance – when it ran in front of an infra-red trail camera, and the records we have would suggest that there are dormice in all Ranscombe Farm's woodlands.

Harder still to trace is the harvest mouse *Micromys minutus*. Like the Dormouse, it is a name which many people will recognise but an animal which very few people will have seen. Harvest mice are tiny, with a head-and body length of just 5-7cm (2-2 ½ inches), and are the smallest species of rodent in Europe, so that doesn't help. We probably wouldn't have known there were harvest mice at Ranscombe Farm at all if it hadn't been for the help of Dr Pam Worrall of Kent Mammal Group, and Steve Kirk who ran the Kent Harvest Mouse Survey. Pam asked Steve to come in and run a harvest mouse training day for the Ranscombe Wildlife Study Group volunteers – and when we went out in the afternoon to try out the field survey method, we ended up finding three harvest mouse nests! It turns out that the best way to check for the presence of harvest mice is to go out in early winter and look for abandoned nests – these are made from blades of grass, which are carefully split and then woven together into a ball, around the size of a tennis ball. The nest is built around the stiff flower-stems of tall grasses so that it is held suspended around 30-60cm (1 or 2 feet) above the ground. Harvest mice are great climbers, and spend most of their lives off the ground and clambering amongst grasses and other vegetation, where they are aided by their small size and their long, prehensile tails which can grip onto plant stems like a fifth limb.

Repeated surveys in subsequent years have generally found just a small number of harvest mouse nests, but in different parts of the reserve and often on the edge of a field which was growing wheat or barley over the previous summer. It is this association with cereal crops that earned the harvest mouse its name, as traditional cereal crops suited them well – before the development of modern crop varieties, wheat and barley would be long-stemmed and stand maybe a metre (3 feet) tall and also be harvested later than modern crops. Add to that the presence of more weeds (and therefore seeds) in the era before herbicides, and many more insects amongst the crop for the mice to feed on, and it would have been a perfect home until the disturbance of harvest-time. But even then, pre-industrial methods of farming seemed to suit them: the cut cereal would be bound up into sheaves, which were carried into the yard and stacked in large ricks, the ricks being roofed with thatch to keep the crop dry until it was needed. The first person to describe and provide a written record of the harvest mouse was the 18th Century naturalist, Gilbert White of Selborne in Hampshire: prior to his writing in the 1760s, it was a species unknown to science, even though it was clearly very familiar to farmers and farm-workers. In The Natural History of Selborne, White writes, "As to the small mice, I have further to remark … A neighbour housed an oat-rick lately, under the thatch of which were assembled near an hundred … I suppose they are the smallest quadrupeds in this island."

With an animal like the harvest mouse, which has such an apparent association with farming and farmed landscapes, you have to ask how and when they got to Britain. If most cornfield weeds came with the early farmers, then might the same be true for harvest mice? Were they already here when people first arrived in the Mesolithic, and, if so, where did they live before there were cereal fields? Or were they accidentally brought in by early farmers, hidden amongst sheaves of wheat or barley, in the same way that animals may have found their way into corn ricks in Gilbert White's time? It seems that a lack of fossil evidence means that nobody really knows when harvest mice arrived, but it is the case that they are not completely tied to farmland and do make use of other habitats, particularly wetland habitats like reedbeds. One study in Suffolk found a greater number of nests in reed-beds, fens and marshes than in arable field margins, though this may perhaps have something to do with the earlier harvesting of modern winter-sown cereals compared to the 18th Century, and there being no modern equivalent of the warm, dry winter homes provided by old-fashioned corn-ricks.

But it could be that harvest mice were originally specialists in wetland habitats and tall herbaceous vegetation which found their own way to Britain and only spread into cornfields much later. It is possible to imagine that, even if Britain was largely forested by the end of the Mesolithic period, there would be places along river valleys or around the coast with plenty of open wetland. But what about some of the birds we associate with farmland? Were they here already or did they follow the early Neolithic farmers? Yellowhammers have a strong association with hedgerows and scrub (I suspect that the absence of yellowhammers at Ranscombe Farm may in part be due to the site having only a few hedges) and some models of the original North European wildwood suggest there was plenty of scrub as a result of the work of large grazing animals such as cattle and horses. But skylarks and corn buntings are much less associated with scrub and hedgerows. They prefer wide-open landscapes, especially skylarks which like to have a good view all around: the website of the British Trust for Ornithology suggests that skylarks probably evolved as birds of steppe grassland and coastal saltmarshes where short, open vegetation would occur naturally.

Steppe is a type of natural, grassy plain that occurs in temperate parts of the world where there is so little rainfall that tall grasses and trees find it very difficult to grow there. The Eurasian Steppe extends from Hungary all the way to China, so skylarks could easily have spread westward as farming moved out into Europe from the Fertile Crescent – indeed, modern maps show that Central Eurasian skylarks migrate southwards for the winter, so perhaps this is how they first encountered the early farmland of the Middle East and found it to their liking. Corn buntings, too, are found across Europe and into Asia and North Africa, so maybe they also followed farming through Europe as humans opened and cultivated the landscape.

We are used to the idea that the landscape of Britain has been – and continues to be – massively altered by human activity. But perhaps we see our impact as largely being to mess around with a set of species that were already here when humans first arrived, knocking out some and causing others to increase. Well, we certainly did that, but as with the cornfield weeds I talked about in previous chapters, it could well be that our impact has been greater than that, and that farming brought with it its own whole suite of species – a human-centred ecosystem, if you like – that travelled with farming people and farming culture and brought an entire new component – and new species – to these islands.

Previously, I have mentioned how the common poppy may have actually evolved in cultivation, rather than being an existing wild plant that moved into farmed fields, and I argued that perhaps interrupted brome might have done the same. Well, while reading up on farmland birds for this chapter, I came across work that suggests that the house sparrow *Passer domesticus* – one of our most familiar birds, in both town and country – probably also evolved as a new species in response to early farming in the Middle East. DNA evidence suggests that right at the start of modern farming around 10,000 years ago, an existing sparrow species started to split into two distinct lines: one which kept away from humans and had a more slender bill for eating wild grass seeds, and another that lived alongside people and had a larger, stronger bill for eating cultivated grains. The latter is our house sparrow, which was then able to spread through

Europe with the spread of farming. House sparrows even seem to have evolved a special adaptation to eating a high-starch diet of cereal grains, having particular genes which code for the enzymes that break down starches; similar genes also appeared in Neolithic humans as a probable result of taking up the farming lifestyle.

It may seem a bit of a jump from talking about some of the birds and mammals of Ranscombe Farm Reserve, to discussing the evolutionary origins of the house sparrow. But if you are concerned about conserving wildlife, it is important to try and understand why a particular species is in a particular place and doing whatever it is it is doing. This is especially true where a species is considered to be under threat: in the case of birds, this means being on the Red or Amber Lists of Birds of Conservation Concern which is published by a group of organisations including RSPB, the British Trust for Ornithology, the Game and Wildlife Conservation Trust and others. The lists take into account changes in the size and range of populations, overall population size and other issues such as whether the UK population of a particular species is of international importance. House sparrow, skylark, corn bunting, yellowhammer and turtle dove are all on the Red List because of long-term declines in their populations.

All of these species are considered to have declined because of 20th Century changes to the way the land is farmed. Yet the numbers of all these farmland species – harvest mice too – must have originally been higher only because land was being farmed in the first place. For the house sparrow, the species may not have even existed if it had not been for farming. Other species may have seen numbers rise and fall because of changes in farming practices, or even because there were economic changes which favoured one kind of crop over another. As just one example, the Medway-based birder, Kevin Thornton, recently posted an interesting thread on Twitter where he noted – drawing on old reference works – that turtle doves were considered scarce birds until a rapid increase from the mid-19th Century onwards, probably in response to an increase in the amount of arable farmland This doesn't mean that the current and very severe decline in turtle dove numbers is not a cause for concern, but that there is an important context to this decline which needs to be understood if conservation is to be successful.

I would emphasise, given how much I have written about the wonders of farmland wildlife, that I am not one of those contrarians who will tell you farming has done all manner of great things for the natural environment. Farming, as a basis for modern society in Britain, has most certainly had a hugely negative impact on the islands' biodiversity, in spite of the gains for some species. Anyone who thinks that farming is automatically good for wildlife or the landscape is greatly mistaken – apart from anything else, that's not the point of farming, which is there to grow food. But that doesn't mean that blame should be laid at the farmers' doors. Farmers are just like everyone else: a few are greedy and uncaring, a few are totally selfless, and most are honest and hardworking people trying to do a good job – they want to be able to grow good food, to leave the environment in a decent state, and to make a living in difficult circumstances (and the economic difficulties suffered by many farmers should not be underestimated). We all need the food that farmers produce, so it is no good if us non-farmers have a go at them for doing it efficiently. I may be too heavily influenced by the 18th/19th Century writer and polemicist William Cobbett (something of a personal hero, though I would acknowledge his substantial flaws), who was a great champion of the small farmer and who reckoned bread, bacon and beer was the best diet for a working person – but then I really like good bread, beer and cheese and only a farmer is going to be able to put that on my table for me. If we want good food at an affordable price as well as a healthy environment, then it takes all of us together – food producers and those that only consume – to work together.

The impact of larger mammals on Ranscombe Farm Reserve's other wildlife is very hard to gauge. Badgers shift a lot of soil, and there have been a couple of hints of their impact at Ranscombe Farm. In one location on the very edge of an arable field, the digging of a new entrance hole in an established badger sett produced a big pile of bare soil and an appearance by rough mallow amongst what was otherwise long, dense grass and brambles. Badgers may also be responsible for one of the reserve's small colonies of lady orchids. By bringing up chalky subsoil and spreading it over the clayey surface soil, they may have changed the soil drainage and chemistry – making it drier and more alkaline – by enough to suit these somewhat fussy plants; certainly all the lady orchids in this particular colony are growing in and around the diggings of a long-established sett. Interestingly, research in a Polish forest also found badgers reduced the acidity of the topsoil, as well as noting that badgers' habit of eating fruit meant a lot of pips and stones were passed out in their dung, leading to an increase in the diversity of fruit-bearing trees and shrubs.

Seed is not just carried in dung, of course, and wandering badgers must help to move plant seeds about the place, either attached to their fur or in mud on their feet, as, presumably do the red foxes *Vulpes vulpes* that make a home at Ranscombe Farm. Most of this movement must be fairly local though, as both badgers and foxes spend most of their time within an established home range, which research by Oxford University's Wildlife Conservation Research Unit (WildCRU) found to generally be somewhere around 30 to 90 hectares (about 75 to 225 acres) for badgers, and to average around 40 hectares (around 100 acres) for red foxes in the areas they studied. Yet the amount of animal movement, and therefore the potential amount of seed movement within these localised areas must be quite substantial, given that maybe up to a dozen badgers might be active in a single home range (WildCRU found their study area held an average of 40 badgers per square kilometre), and that 4 or 5 adult foxes may share their 40 hectare territory.

The cattle which graze the grassland at Ranscombe Farm are probably providing a similar seed-transport process, but even the largest area within which cattle are able to move freely is less than 14 hectares (34 acres). And, unlike badgers or red foxes, the cattle are unable to move between woodland, grassland and arable, which means that they are mainly stirring an existing pot and are less likely to add new ingredients. The animals likely to be the widest-ranging travellers at Ranscombe Farm Reserve are probably the fallow deer which move freely across the reserve and are often spotted during the day, sometimes in groups of maybe eight or ten animals – occasionally even more – and in a variety of colours from dark to completely white individuals. The total number present on the reserve is very hard to gauge, as the deer keep to the woods for most of the time, and they are likely to be roaming over a much wider area than the reserve itself.

Deer are known to carry wild plant seeds both inside and out, with one German study finding that roe deer carried seeds of 38 plant species on their fur and hoofs, with seeds of just seven species in their dung, while some work done in Eastern England found seeds of 33 plant species in the droppings of red deer *Cervus elaphus* and fallow deer. The one case of significant plant movement at Ranscombe Farm for which a fallow deer must be the most likely suspect was the unexpected appearance of broad-leaved cudweed on a footpath more than 200m (220 yards) into Cobham Woods, well away from its normal arable habitat. The seed from which this plant grew is likely to have been picked up on the hoof of a deer which had wandered across an arable field, hopped the fence, and taken the path up through the woods. It is possible that the real culprit might instead have been a human visitor's boot, but this would have meant the seed being carried travelling more than twice the distance, as people do not take to leaping over a 4½ foot (135cm) wire fence quite as casually as a fallow deer does.

Fallow deer are a mixed blessing. They are useful in moving seed about, and large herbivores, including deer, are considered an important part of rewilding projects. But deer are often seen as a problem in woodland management, and in places where deer numbers are high, they can cause serious damage by eating tree seedlings and the young stems of recently coppiced trees. Fortunately, at least so far, this has not been a problem at Ranscombe Farm, perhaps because deer numbers are relatively low. This does not mean that there is no damage to young or regrowing trees, and it is certainly possible to spot where young tree shoots have been nipped off: often this is alongside a path, where animals have just stopped for a quick snack while on their way to somewhere else. But the experience at Ranscombe Farm has been that the rapid regrowth of coppiced trees very soon outpaces any deer nibbling. Any real deer damage seems to be more localised – for example, they certainly seem to like the taste of lady orchid flowers, so any plants which are obviously visible do tend to get nipped off, and this must be having an impact on our lady orchid populations. Male fallow deer will also damage trees when growing their new antlers each year: to get the dead skin off the developing antlers, they thrash them against young trees which results in the trees losing bark and branches. The trees which the deer select for this punishment always seem to be ones that have been deliberately retained for conservation purposes, so that ensuring the survival of new field-edge oak trees has meant equipping them with their own miniature timber enclosures. Ranscombe Farm's fallow deer are also no friends of farming, often being found out in the arable field where they like to graze on the growing crops, and they have even flattened large patches of crops after choosing the middle of a sunny field as a suitable place to lie up.

But the fallow deer of Ranscombe Farm Reserve are also significant as an illustration of one of the less obvious ways in which human behaviour has influenced the British landscape and its biodiversity – for good and bad – and that is through political structures. Although fallow deer were apparently present across Europe before the last Ice Age, they never made it back to Britain after the ice retreated. The fact that they are here now is thought to be due to the Normans, who introduced fallow deer to Britain for hunting in their parks and forests: the Normans loved to hunt and of course it was William the Conqueror who created the New Forest as his own royal hunting preserve, back in 1079. Ranscombe Farm's fallow deer probably originate from the deer park of next-door Cobham Hall, which seems to have originally been set out in the 16th Century by the tenth Baron Cobham, William Brooke. William Brooke was a descendant of Henry de Cobham who bought the Manor of Cobham in 1208, and who, according to a 19th Century edition of the journal *Archaeologia Cantiana*, may have been one of the crusaders present at the siege of Acre (in modern-day Israel) in 1191. The whole family lineage from Henry to William had close ties with the Crown, with various members serving as Sheriffs of Kent, and so in charge of law and justice in the county, and one of William's antecedents was with Henry VIII at the sumptuous summit with the French King known as the Field of the Cloth of Gold. William himself held the title of Lord Warden of the Cinque Ports and entertained Queen Elizabeth I at his home at Cobham Hall.

Henry de Cobham's descendants were serious political players, and it is through politics that they held onto their estate and their influence. Cobham Hall and its grounds have had a long and complex history in the 500 years since William Brooke was in residence, and, as I described at the start of this book, the house and land have been variously bought, sold, inherited and gifted to different people during this period. The human lineage has been broken several times between Henry de Cobham's purchase of the Manor of Cobham back in 1208, but the fallow deer you see at Ranscombe Farm Reserve are an outcome of – and a direct though distant link to – the power and politics of the Middle Ages.

Chapter 11 - Managing to be wild

Most first-time visitors to the site may well be surprised that Ranscombe Farm Reserve remains a working farm, with a commercial tenant farmer. About 80 hectares (197 acres) – getting on for a third of the whole site – is managed as arable farmland, with (at the time of writing) about 25 hectares (62 acres) of this being in large blocks which are dedicated to conserving cornfield weeds. That leaves some 55 hectares which continue to turn out a commercial crop, which is normally wheat, barley or oil-seed rape but might include field beans or flax. These cropped fields are far from being conventional, however. Broad, uncropped margins are left around each of them: in some places, these margins have good populations of cornfield weeds and are turned-over each year; in others, they have been left to turn naturally to grassland or scrub, or both.

But even away from the edges, deep in the crop, the arable fields at Ranscombe Farm are far from average. The farmer at the reserve, and farming the site long before it ever became a nature reserve, is Andrew Lingham, one of many farmers now taking up the baton of 'regenerative agriculture'. Broadly, the aim of regenerative agriculture is to work with nature and to improve the environment, while still remaining productive and commercially viable, and it takes into account things like soil health, carbon emissions, and flooding and water management.

For instance, Andrew's approach to growing arable crops involves no ploughing or other cultivation of the ground: so-called no-till farming or direct-drilling. For centuries, the way to create a seedbed for sowing a crop started with using a plough to turn over the soil surface and bury any existing vegetation. Then you would run a harrow over the ploughed ground to break up the big clods of earth to create a fine soil surface. Then you sowed the seed and harrowed and rolled the ground to push the seed into the soil. After Jethro Tull invented the seed drill at the start of the 18th Century, seed was directly planted in rows by machinery towed by horses and later by tractors. The no-till method differs from this in that it avoids disturbing the ground, and just skips to the final stage of drilling seed into the soil. There is no ploughing or harrowing. And, while the process does often mean using herbicides to kill off any weeds prior to drilling the seed, there is the (by no means small) compensation of a massive reduction in the use of diesel, as the tractor just needs to make one pass over the field to do the whole job. Herbicide use can also be reduced if a cover crop, such as field beans, is used whenever the soil might otherwise be left bare between harvest and sowing.

Arable cultivation is an important part of the management at Ranscombe Farm Reserve. Photograph: R I Moyse/Plantlife.

Leaving the soil undisturbed leads to a big increase in the amount of organic material it contains, locking up carbon that might otherwise be lost to the atmosphere. Repeated ploughing tends to do the opposite, exposing buried organic matter to the air, where it oxidises and adds to the carbon dioxide in the atmosphere. Under no-till, there is also an increase in the life in the soil: without their home being ripped-up every year, earthworms, ground beetles and other invertebrates can build up much bigger populations and become the farmer's allies. Earthworms are good at incorporating organic matter into soil, and play an important role in nutrient cycling; many ground beetles are predators of pests such as slugs. A bit of work by Dr Simon Springate of the University of Greenwich showed just how well Andrew's no-till approach has worked for ground beetles. In a comparison between arable fields at Ranscombe Farm Reserve and similar fields on a nearby organic farm and a conventionally managed farm, Simon found that the ground-beetle populations at Ranscombe Farm were the largest by a very long way (though the organic farm proved best for spiders, which are also important predators in crop fields).

Cattle are an important part of the farming system at Ranscombe Farm Reserve. Livestock farming, and especially cattle farming, has had a bad a press in recent years for the contribution it makes to global warming. But without grazing animals, it would be impossible to restore and recreate wildflower meadows at Ranscombe Farm Reserve, especially the substantial 23 hectares (some 60 acres) of Brockles Field. Even if mowing did the same job (and it does not: mown grasslands are distinctly different to grazed grasslands), it is hugely expensive to cut a big field like Brockles and remove all the cuttings, and it relies on powered machinery; cattle will do it for free. So having some kind of large grazing animals at Ranscombe Farm Reserve was always going to be essential, and, because Andrew Lingham has always run livestock, it was a natural fit to combine the conservation grazing with Andrew's farming. The cattle which Andrew uses are hardy animals which are able to do well on natural forage and do not need to be fed on grain in order to put on weight. This means they can spend time grazing the wildflower meadows and getting these important habitats into good shape, and also go onto some of the commercially-farmed arable land in winter, where they graze on a sown cover crop. This winter grazing on the arable helps improve the soil as the animals' heavy hooves tread some of the cover crop down into the ground, and fresh cattle dung adds nutrients. The result of all this useful activity by the cattle is another source of income for Andrew's farm business, producing beef that is sold through his small farm shop in the nearby village of Halling.

The value of good dung is not to be underestimated, and not just as fertiliser. The beetles found in fresh cattle dung by intrepid (and gloved) Ranscombe Wildlife Study Group volunteers ranged from the big, shiny dor beetle *Geotrupes spiniger*, nearly an inch (more than 2cm) in length and half-an-inch (12mm) across, to the tiny *Cercyon pygameus* (barely more than 1mm in length) and included plenty of the black-and-brown beetles in the genus *Aphodius* which make up part of the diet of the threatened serotine bat *Eptesicus serotinus*. Also making good use of cattle dung at Ranscombe is the contender for the title of Britain's largest fly. The hornet robber-fly *Asilus crabroniformis* is big, getting to 25mm (1 inch) or more in length, harmless to humans, but a fierce predator of other insects (not hornets, though – the fly is so named because it is brown and yellow to mimic a hornet and so dissuade larger predators from picking on it). It is a rare insect in Britain and in the South-east it is only found in scattered locations. But it has had an association with the East Thames Corridor, presumably because of the continued grazing of the marshes: the fly's maggot larvae live in dung – especially cow dung – where they hunt larvae of other dung-living insects. Threats to this impressive creature come from abandonment of grazing but particularly from the use of modern worming compounds, like ivermectin, which travel through the whole system of the treated animal and continue to work in the dung, which then itself becomes toxic to insects. Andrew Lingham's regenerative farming approach means that his cattle are not routinely treated with systemic wormers like ivermectin, and so insects are able to get to work quickly and the dung is rapidly incorporated into the soil.

The hornet robber-fly is dependent on fresh, untreated dung. Adult flies hunt other insects - this one has caught a bush-cricket.

Not every aspect of regenerative farming fits neatly with the conservation of farmland wildlife. Most notable at Ranscombe Farm Reserve is the fact that no-till arable farming means that the buried seeds of cornfield weeds do not get brought up to the soil surface. For the arable farmer, of course, the fewer weeds the better. But it does not help with the conservation of rare plants of arable farmland, as their seeds stay buried and never get a chance to grow. If poppies and other cornfield weeds are to do their stuff, then a good going-over with a disc-harrow is needed, if not an actual plough, which means that a lot of the nature conservation work at Ranscombe Farm Reserve probably does more damage to the soil (though over a relatively small area) than the commercial farming. So the no-till approach means no big show of cornfield flowers, but it also means the bank of weed seeds in the soil is likely to remain intact and in place, ready to take advantage of future opportunities; this contrasts with conventional farming, where cultivation encourages cornfield weeds to germinate, only for them to be killed with herbicide, so that the seed bank in the soil may become rapidly diminished.

Plantlife could have chosen another approach when they took on Ranscombe Farm Reserve, and the site did not necessarily need to remain as a commercial farm. But the advantage of working with a real farmer in the real commercial world is that you can learn lessons about how to make farming work for plant conservation, especially on a site as important for cornfield flowers as Ranscombe Farm. One of the most important lessons was just how hard it is to do conservation for rare cornfield weeds.

In theory, it ought to be quite simple to get a field full of poppies and other arable wildflowers, provided the seeds are still present in the soil. Plough and harrow, perhaps sow lightly with some crop seeds, then walk away. This is the basis of those various options in UK agri-environment schemes which are targeted at rare cornfield weeds – the kind of work which farmers are encouraged to undertake in return for compensatory payments. It does work, and at Ranscombe, we saw populations of rare plants pop up out of the soil and then spread and increase. But it has to be done right, and timing, for instance, appears to be important. Some cornfield weeds germinate best in the spring – fumitories, for instance – and really want any cultivation to be carried out then. Others respond best to autumn cultivation, and this needs to be done pretty early in the autumn as well: autumn cultivation and sowing for crops may be carried out as late as November, but this seems to be too late for some autumn-germinating plants, like prickly poppy, which may not be able to get going in the colder weather.

Get the timing right, though, and you can give the rarer plants the best chance to germinate and flower and set seed. But keep repeating this annual cultivation without a break, and the vegetation quickly changes. At Ranscombe Farm Reserve, we saw the appearance of big populations of annual grasses including rat's-tail fescue *Vulpia myuros*, soft-brome, barren brome *Anisantha sterilis* and great brome *Anisantha diandra*, so that arable fields could quickly come to resemble tall, grassy meadows. Some biennial and perennial plants also managed to flourish despite the fields being thoroughly harrowed and broken up to produce a completely bare soil surface. The tall, blue-flowered spikes of viper's-bugloss can sometimes colour the whole of Longhoes Field in high summer: it is a plant loved by bees and butterflies and normally welcome, but here it takes up space that could be used by some much rare species. In Kitchen Field, the tall stems of perennial sowthistle *Sonchus arvensis* swept across most of the field for a year or two, carrying their big, beautiful flowerheads which look like golden pennies, and then were replaced by two other yellow-flowered members of the daisy family, bristly oxtongue *Helminthotheca echioides* and hawkweed oxtongue *Picris hieracioides*.

The unexpected abundance of these plants can often mean that the rarer cornfield plants get elbowed aside and pushed into second place in the race to fill the available arable space – this is no worse than having to fight for a space in a wheat or barley crop, perhaps, but not the best possible result. Fortunately, we found that in Kitchen Field we could keep the oxtongue populations from overwhelming the whole field by letting cattle have a go

Viper's-bugloss populations can build up with repeated cultivation of arable fields, as here in Longhoes. Photograph: R I Moyse/Plantlife.

at them in autumn. Despite the very rough and hairy leaves which give oxtongues their name, the cattle find them extremely tasty, and will happily munch up any they can find, and this does seem to help. What we were not able to do was to use the old-fashioned way of dealing with a build-up of weeds, which would be to put the field down to closely mown or grazed grass for a couple of years before ploughing it again. This approach would have been impractical for reasons of time and cost, but would also have meant that corncockle and perhaps interrupted brome, with their short-lived seeds, would not have survived without regular, deliberate sowing.

There is more that needs to be learned about how rare cornfield weeds can best be conserved as part of the farmed landscape – or, indeed, in the wider landscape. But if Ranscombe Farm Reserve can offer one clear lesson about how to get wildlife back into the farmed landscape, then that lesson is to give nature the space it needs and then to trust the wild plants and animals to get on with the job. One of the best ways to make that space available is to deliberately blur the boundaries between woodland and open fields, so that instead of just two habitats – woodland and grassland, or woodland and arable – there is a whole mix of mini-habitats in between, including scrub, tall grass, short grass, and perhaps some uncropped bare or cultivated ground. A lot of wild species love to occupy this middle ground, and it is these scraggy, indefinite margins that are some of the richest and most interesting bits of Ranscombe Farm. Take two or three steps one way, and you can be with the primroses and lesser celandines under the shady woodland edge; head in the opposite direction and it is poppies and clovers and mayweeds. Between the two, amongst the tall grass and brambles might be small scabious, greater knapweed, or common gromwell, while low-growing common rock-rose and long-stalked crane's-bill might be on the very fringe of the arable. The complex and intricate structure of these graded margins and corners makes them hugely attractive to animals, whether these are the glow-worms *Lampyris noctiluca* that hang in the tall grass on a summer night, the viviparous lizards that climb up onto bracken to bask in

Bristle club-rush. Photograph: R I Moyse/ Plantlife.

spring sunshine, the whitethroats that build their nests low amongst the bramble, or the harvest mice that retreat into the tall field-edge vegetation after the crop is cut and gathered.

Making space for nature also means not being tidy. A bit of scruffed-up, turned-over or disregarded ground may be just the opportunity some plant or insect is looking for, like the common eyebrights using bare scrapes as their entry points into Brockles Field. In the woods at Ranscombe Farm, the deep wheel ruts left by forestry vehicles provide standing-water habitats for plants like brooklime and water-pepper *Persicaria hydropiper*, hold water beetles and pond-skaters, and are sometimes home to tadpoles. The only places where the little, grass-like bristle club-rush *Isolepis setacea* has been found at Ranscombe Farm is along the edges of woodland tracks where the ground gets scuffed as the big forestry machines slide on the wet track surfaces. Yet some forestry management standards look for wheel ruts to be filled in once a season's work has finished, leaving the tracks neat but comparatively lifeless. Likewise, certain land management schemes aim for neatness and consistency, and frown on having scrub in wildflower meadows, even though scrub can add to plant diversity – burnet rose and small-flowered sweet-briar *Rosa micrantha* add to the sights and scents of chalk grassland at Ranscombe Farm Reserve – and create shelter, cover and structure for insects and other animals.

The experience at Ranscombe Farm also makes the case for trusting nature to take the opportunities offered. It might take time, but wild plants have a knack of finding their way in and setting up home. Brockles Field has become a major piece of wildflower meadow because the plants were able to do most of the work: management of the field just used mowing and grazing to keep it as grassland, and all the plants – from oxeye daisy *Leucanthemum vulgare* and grass vetchling to large thyme and man orchid – found their own way in. Longhoes Field was a conventional, largely flower-free arable field till a change in management gave the cornfield weeds a chance, and now must rank as one of the richest arable fields in Britain. In both cases, a result which was at least visually similar might have been achieved by buying and sowing commercially available seed, but doing so would have reduced the amount of space in which nature could be itself. There are plenty of times and places where buying and sowing seed might be the only option, of course, and Ranscombe Farm Reserve was always in the lucky position of having some excellent flower-filled patches right from the start – and even then, some plants seem unable to move far without assistance. Yet it is still a good rule to interfere as little as possible in nature's own processes of habitat creation, and at least to give it the chance of showing what it can do unaided and what improvements it can make to your own efforts.

More than just scrub - small-flowered sweet-briar in the grassland above Kitchen Field. Photograph: R I Moyse/Plantlife.

This applies to woodlands too. Planting new woodland is popular and generally well-intentioned, but if the aim to is to make new woodland to benefit wildlife, then planting is usually entirely unnecessary. Most trees are very good at sending seed out into the world, and all that is usually needed to turn a patch of land into woodland is to leave it alone. This is not only easier and cheaper than tree-planting, but gives nature the chance to decide what species to put in, and makes space for all the plants and animals of scrub and early woodland succession. Something very similar applies to existing woodland. Ben Sweeney, the warden at Ranscombe Farm Reserve, has been able to recruit wild plants to start turning the near monoculture of sweet chestnut in Cobham Woods into something more mixed. Although labour-intensive, the process is pretty simple: where blocks of woodland are coppiced, or strips cut to produce scrubby margins along the open rides, the regrowth of the sweet chestnut is suppressed by cutting or the targeted use of herbicides. This gives other, slower-growing trees and seedlings – in Cobham Woods, this includes pedunculate oak, hawthorn *Crataegus monogyna*, silver birch *Betula pendula*, hazel, holly *Ilex aquifolium*, and goat willow *Salix caprea* – a chance to get away without being smothered by shade within a couple of years.

You may feel that there is an echo of rewilding in these ideas about making space for nature to act in its own right. Yet it does seem unlikely that Ranscombe Farm Reserve is ever going to be a truly wild place. Its most important species are the rare and vanishing weeds of arable farmland – which might almost be considered the exact opposite of a wild habitat – and these plants alone are enough to justify the site's existence as a nature reserve. But even with the conservation of cornfield weeds at Ranscombe Farm, and much more so with the work on grassland and woodland, there are aspects of the reserve's more 'traditional' habitat conservation work which have something to say about rewilding and how it might work.

Like 'wildlife' and 'wildflower meadow', 'rewilding' is a term that is widely used but which does not seem to have a universally understood meaning. I am not going to try and add my own ideas about what rewilding really means, but most of the definitions I have come across, online or in printed articles, talk about the restoration of active natural processes to create conditions which support wildlife: such processes might include ecological succession, grazing by large herbivores (and perhaps also predation of those herbivores by large carnivores), flooding, erosion and other factors. Working at a large scale also appears to be important. This

Traditional conservation in action: volunteers cutting and raking the grassland above Kitchen Field. Photograph: Ben Sweeney/Plantlife.

does seem fairly clear as a big, over-arching idea, though it runs into the problem that we do not know what natural processes might be needed by a wild ecosystem, or whether they could all be reinstated in practice. But I do have serious problems with the idea, sometimes stated in definitions of rewilding, that "nature knows best". As a statement, this could perhaps be said to stand up as a matter of faith, but from a purely scientific viewpoint, nature knows nothing and neither does it care. The history of the world shows that, for all that it has given us this amazing, wonderful world, nature has a terrible record on extinctions, with five mass extinction events up to and including the one that wiped out the dinosaurs. However much we may see nature as a source of solace or wonder, or draw lessons from it, it has no will or direction of its own: without wanting to go too far down a philosophical rabbit-hole, the only judges of whether or not nature is doing a good job are us humans.

Amongst the ways in which rewilding is defined, there also seems to be an assumption that rewilding will automatically lead to an increase in biodiversity and, possibly because of this, it does not need a defined end-point: we just need to set off the rewilding process and let it run. In practice, it is hard to see why this should necessarily be true in every case, and the impact of rewilding on the overall biodiversity of a given area seems likely to depend on the scale of the project and its starting point. After all, 'traditional' conservation methods are largely versions of less intensive farming and other economic land management, which include cutting or grazing of grassland and heathland, coppicing of woodland, and mowing reed-beds, and the experience of the 20th Century has taught us that the countryside held more wildlife when it was all managed in this way. Apply these traditional techniques at the kind of scale envisaged by the rewilding movement and you might well end up with more biodiversity than a truly wild landscape would hold. I strongly suspect that the wild Britain encountered by the first Mesolithic arrivals contained fewer species than were present here later, after humans started to have a major impact, and perhaps fewer even than today. This does not, though, mean that a less biodiverse result would necessarily be worse: it is entirely possible to envisage a situation where rewilding results in there being fewer species overall but much more positive results for those species – and there are many, many of them – which are still declining despite a century or more of conservation action. In global terms, an increase in those habitats that are particular to Britain – especially temperate rain-forests of the western coasts and some marine and coastal habitats – could be a significant conservation win, even if it was associated with losses of species and habitats that are more widespread across Europe[1].

Some of the support for the current drive towards rewilding is that so-called 'traditional' conservation methods have failed. This is by no means true, however – or, at least, not wholly so. Ranscombe Farm Reserve has achieved a lot within its boundaries for rare plants, and this has meant that rare animals have also benefited; Ranscombe Farm is not exceptional in this regard and there are many, many other nature reserves where the successes are obvious and the same could be said. Beyond individual sites, many past conservation efforts have proved very successful where the causes of the disappearances of plants and animals could be tackled through legislation. For example, the big successes, especially with birds of prey, seem to be where the original declines and disappearances were down to a combination of persecution and persistent pesticides. It was because these issues were successfully tackled that we now have buzzards, peregrines, sparrowhawks and ravens at Ranscombe Farm Reserve. Other species groups often tell a similar story to wild birds. For example, the conservation charity Butterfly Conservation has reported that three-quarters of the UK's

[1] When botanists from South-east Europe visited Ranscombe Farm Reserve as part of Plantlife's Natural Networks project, they expressed surprise at our focus on meadow clary – to them a relatively common plant – but were excited to see primroses. How we see conservation priorities is dependent on the scope and scale of our view.

butterflies are in decline, and that total moth numbers have dropped by more than a quarter since 1968. Yet, as with birds, where it has been possible to carry out targeted work, vanishing species have been brought back from the brink: in East Kent, for instance, grassland restoration and recreation has massively expanded the population of the rather smart-looking and highly threatened black-veined moth *Siona lineata*.

On individual sites and with individual species, 'traditional' conservation has undoubtedly been hugely successful; it has even managed to slow – at least in some cases – the widespread and continued erosion of both the quality and quantity of wildlife habitats across the British landscape. But what it has not yet been able to do successfully is to turn the tide and rebuild biodiversity at a national scale. As with the butterflies, the spectacular successes with birds of prey need to be viewed against a background of steady disappearance: the report, The state of the UK's birds 2020, notes that the total population of native breeding birds today is only four-fifths of the population present in 1960. So there have been major successes for the traditional conservation approach, and perhaps some species' declines have been slowed, but, despite this, biodiversity is still being lost.

With wild plants, detailed monitoring across the wider landscape is not as advanced as for birds and butterflies. Though atlases showing plant distribution are widely available at 10km scale, or 2 km scale in county floras, this does not give a very good feel for population change: a dot on a map may represent one plant or ten thousand. It is only in very recent years that the National Plant Monitoring Scheme has started to study more closely the fortunes of plants in the wider countryside, though research by Plantlife, published in its 2012 report, Our Vanishing Flora, has suggested that the average UK county was losing nearly four species of wild plant every decade and that the rate of loss seemed to be accelerating. Nature reserves, like Ranscombe Farm, and other protected sites can do a great job of holding the line for many plants, or even, as with interrupted brome, provide a base for their re-establishment. But unless we tackle the wider losses, these patches of protected land will increasingly become isolated and besieged fortresses in a sea of ever more impoverished landscape.

So, however cynical I sound, and whatever reservations I have about the way rewilding is presented, there is no doubt in my mind that the rewilding approach will prove vitally important in the future conservation of wildlife in Britain. Not because nature knows any better than you or I, nor because there is anything fundamentally wrong with the way conservation has been practised in Britain up until now. To me it is important firstly because it allows natural habitats to go through a constant process of change, rather than us trying to hold habitats at a particular stage in ecological succession, whether this is grassland, scrub, coppice woodland, open water, heathland or whatever. The subtle gradations in time and space between the various stages of succession are almost certain to provide situations and sets of conditions that we have not even thought of, and the process may also help to provide more very early and very late successional habitats – the bare, brightly-lit, open ground and the deep, dark, old growth woodland – that are very definitely in short supply in this country.

Secondly, rewilding is likely to be the most cost-effective way of restoring wildlife habitats at a very large scale. However well traditional conservation management works, it can be prohibitively costly for conservation charities and governmental bodies to deliver, even with the hugely important contributions made by volunteers. Ranscombe Farm Reserve, at around 250 hectares, runs on a minimum of one member of staff and a dozen or more regular volunteers (not including the supporting staff at Plantlife HQ, or the support from Plantlife's members): scale that up to 5,000 hectares and you would have to multiply those staff and volunteer numbers by twenty to maintain the same intensity of management.

Finally, I think rewilding will surprise us with results we did not expect, and probably could not expect, and that will teach us huge amounts about the way different species behave and respond to the world around them. Because if there is one lesson to be learned from existing rewilding experiments (and however much I've emphasised that nature doesn't know or care about biodiversity), it is that our own knowledge or natures processes and interactions is slight at best. We will only learn what wild plants and animals really need if we let go of our assumptions and give nature its head.

One of the big lessons from the rewilding experiments on the Knepp estate in Sussex was the explosion of purple emperor butterflies *Apatura iris* and their unexpected association with scrub, rather than the woodland that most people assumed was their preferred habitat (it is woodland that they are occasionally seen at Ranscombe). In even the relatively traditional nature conservation environment of Ranscombe Farm Reserve, the plants never stopped surprising me: the sudden appearance of fine-leaved fumitory from long-buried seed; the unexpected arrivals of common eyebright and yellow-rattle from who-knows-where; the fleeting visits of hoary cinquefoil and brackish water-crowfoot. If the Dutch rewilding site of Oostvaardersplassen is any guide, there could be some big bonuses for wild plants. This immense site of around 20 square miles has had its ups and downs, but one of the amazing and completely unexpected results was the appearance of the wetland plant marsh fleawort *Tephroseris palustris*. This ragwort-like plant is listed as Vulnerable in the Netherlands (it has been extinct in the UK for more than a century), yet the population on the bare, wet mud at Oostvaardersplassen became so huge that the sheer bulk of the plants' fluffy seeds as they blew across local roads was said to be a hazard to traffic.

How can we know that rewilding will actually deliver the outcomes we want? It sounds like a silly question, but if the idea is that rewilding will restore natural processes which lead to increased biodiversity, or the restoration of lost biodiversity, then we have to start with some idea of what natural processes might be needed to do the job. This means diving into a bit of ecological theory.

An interesting paper from 2019 in the journal Science identifies three key elements that contribute to rewilding: in the original language of the paper, these are trophic complexity, stochastic disturbances, and dispersal. In more straightforward terms, 'trophic complexity' means the complexity of the food-chain: things eating other things, whether this is rabbits eating grass, foxes eating rabbits, or beetles and fungi recycling fox and rabbit dung. The implication is that where there are more of these sorts of interactions, then the result will be a more diverse range of environmental conditions, not least because species may behave differently in response to the presence of predators or prey: for example, deer may limit their grazing and browsing to areas where wolves cannot creep up on them. 'Stochastic' means 'relating to probability and chance', so stochastic disturbances are those chance (rather than planned) happenings that reset the ecological clock and allow succession to progress again: for instance, something that turns woodland back to grassland, or grassland to bare ground, and so provides opportunities for a different set of species to use the space. 'Dispersal' is more self-evident: the greater the ability of species to move through a landscape, the more opportunities there will be for new areas to be colonised and for plants and animals to take advantage of the opportunities arising from chance disturbance events.

One of the implications of this is that wildness is a not an absolute: you cannot say that one place is wild and another is not wild. After all, in any place where some sort of vegetation is present there will be some sort of food-chain (even if it is very short and simple), some small likelihood of chance disturbance, and some ability of species to move. Rewilding then becomes about increasing the complexity of food-chains between plants and animals and fungi (and the rest), relying less on human intervention and more on chance natural events to restart processes of ecological succession, and making it easier for species to move from place to place. What this does mean, of course, is that any kind of habitat creation or restoration at any scale can be said to be rewilding if it achieves one or more of these outcomes. So you can rewild your garden, but only so far. It is only when you get to really big scales that natural, chance disturbances can take over from deliberate human disturbances, and it is only in big, joined-up networks of sites that long-range dispersal is likely to become possible for most species.

The Plantlife approach to plant conservation is strongly based on a principle of keeping the 'wild' in wildflower. In other words, aiming to create the conditions plants need to increase their populations and spread through the landscape, with as little help from people as possible. Which means a lot of the focus at Ranscombe Farm Reserve was on understanding plant behaviour and the mechanisms by which plants were able to move about the place. In contrast, a lot of the focus (so far) on the achievements of rewilding has been around benefits for animals, rather than plants. I am not immune to this: the way in which the Knepp rewilding project has worked for purple emperors and for turtle doves has been amazing, and when I visited Oostvaardersplassen, I delighted in seeing birds like great white egret *Ardea alba*, willow tit *Poecile montanus* and my first and only bluethroat *Luscinia svecica*. But I suspect that it will be plants and the less mobile invertebrates that will tell us most about the kind of natural processes that will be needed if rewilding is to succeed.

Meadow clary – the plant which kicked off the Ranscombe Farm Reserve journey – seems to have something to say about the way big grazing animals ought to behave in the wild. It is a tricky plant for conservationists because it does not seem to fit very well with typical wildflower meadow management – at least not up here in the north-west corner of its European distribution. Regular annual cutting or grazing creates a grassland that looks pretty much the same from year to year, and it is an approach that suits a lot of wild plants and animals: if it did not, then wildflower meadows would not be as rich and important as they are. But meadow clary seems to be designed for the kinds of more random, intermittent disturbances which make grassland that some years is lush and tall, and in others is short, sparse and dry. Adult meadow clary plants grow tall and have obvious flowerheads which herbivores find tasty and are happy to eat. Yet the seedlings will only germinate where the surrounding vegetation has been eaten short and where there is plenty of bare ground. Seeds germinate quickly in such situations and will manage very well if heavy grazing continues after they germinate.

In fact, a whole meadow clary population will quite happily persist under heavy pressure from herbivores: the flat rosettes of leaves are held close to the ground and resist grazing and trampling. Because the plants can live for several decades, all they need for survival is for the grazing animals to give them a break and let them flower and set seed every few years. They seem adapted to the kind of irregular cycle of heavy grazing, interspersed with periods with little or no grazing, that you can imagine happening in a wild landscape with herds of free-roaming cattle or horses. By contrast, in a modern farmed landscape, and even on nature reserves, grazing is not only in a regular cycle, but persistent, heavy grazing is often seen as damaging to the pasture and to the productivity of the livestock: too much bare ground and problem weeds might start popping up amongst the grass; too little grass and the animals stop putting on weight.

Irregular, occasional but very major ground disturbance of grassland or scrub seems to be what works for ground-pine: its seeds can remain alive and well in the soil for decades, waiting for the ground to be turned over. Carefully managed pasture, where poaching and over-grazing is avoided, is clearly unsuitable, and the plant does not really seem suited to regular, annual disturbance, or it would be more abundant in arable systems (despite our best efforts, it refuses to appear consistently in arable at Ranscombe Farm Reserve). Although you can find ground-pine on sites where grassland is dug-over repeatedly by rabbits, the thin, poor soil often only produces tiny plants. But turn over ground which has been under grass, and even scrub, for some years – so that a good depth of rich soil has built up – and ground-pine produces big, lush, branching plants: when hawthorn scrub was cleared (actually torn out by the roots) at Trosley Country Park, on the chalk scarp a few miles from Ranscombe Farm, one result was the appearance of ground-pine plants the size of dinner plates. Rough mallow behaves in a similar way, apparently doing better – certainly making bigger plants – where grassland and scrub edges are periodically disturbed, but generally producing smaller plants where ground is more regularly dug over or cultivated.

Some other very rare plants seem to want repeated, very heavy grazing and ground disturbance of a type that would be considered bad farming practice, and which might also be avoided by nature reserve managers: apart from anything else, agri-environment grant payments might be stopped if the land was regarded by the government's Rural Payments Agency as not being in kept in good agricultural condition. Red star-thistle *Centaurea calcitrapa* is one of these plants: it is a very spiky biennial which grows to form rounded clumps dotted with red flowerheads, each head surrounded by stiff, sharp spines, and it seems to do best where the ground is absolutely hammered by heavy grazing animals. A bit like meadow clary, the seeds need to germinate in bare ground, after which they grow to form flat, trampling-resistant leaf-rosettes. In the second year, the plants send up their flowering shoots, and these are so astonishingly tough and prickly that no self-respecting herbivore would go near them; after setting seed, these adult plants then die. Red star-thistle has never been at Ranscombe Farm, but once used to thrive on the open area between Chatham and Gillingham called the Great Lines (and which is just visible from Ranscombe Farm). During my time working at Gillingham Borough Council, I was for a while responsible for management of this area, and saw how red star-thistle did especially well in a small, fenced area where local travellers kept some of their horses: having several animals in a such a small enclosure meant the ground was virtually bare ground for most of the year, and red star-thistle thrived. The council eventually had the horses removed, and the red star-thistle population on the Great Lines consequently dwindled and vanished. Fortunately, Kent Wildlife Trust has been able to maintain a small population not far away on its Darland Banks nature reserve.

Red star-thistle, grows on Kent Wildlife Trust's Darland Banks reserve in Medway. It thrives under heavy pressure from large grazing animals.

Although red star-thistle is an archaeophyte – an ancient introduction to Britain – the way in which many native plants behave would suggest that the wild Britain of the Mesolithic must have included places where grazing and ground disturbance was varied from relaxed to excessive, and others where it may have been heavy and sustained for years at a time. Creating these sorts of conditions in new rewilding projects is likely to be important but a genuine challenge. The land at Knepp is grazed by cattle, but these are reared for market and so need to put on weight and keep it: this means that the grass is not allowed to get so short that the animals run low on food (or at least this was the case when a group of Plantlife staff visited Knepp a few years back). Instead, free-roaming pigs are used to create areas of bare ground. Conversely, at the huge, Dutch rewilding site of Oostvaardersplassen, an unwillingness to interfere with stocking levels led to all the dry ground being reduced to short grass, even previously scrubby areas (the scrub where I saw my bluethroat and willow tit now appears to have completely gone). The result of this was that large numbers of horses, cattle and deer simply starved. It may be that large-scale starvation of grazing animals, perhaps after a year or two of particularly dry or cold weather, is a part of truly wild systems and one of the ways in which ecological succession can restart, but it hardly needs saying that it raises a lot of very valid animal welfare issues.

Random disturbances that reset the clock of ecological succession are not just beneficial to plants that require the occasional burst of bare ground. The fly orchids and lady orchids of the Mill Hill area of Ranscombe Farm Reserve support the idea that there are species ready to take advantage of every step in the succession from open ground to thick woodland. The reserve's biggest population of fly orchids – at its peak maybe 90 plants – is in woodland which is only a couple of centuries old and which was almost totally felled in the great storm of 1987. Fly orchid has proved a tricky prospect for traditional conservation methods: open up the area too much and the plants get eaten by rabbits; let it get too shady and flowering is reduced; and there also needs to be open, sunny ground nearby for its pollinator to nest in. The traditional practice of coppicing woodland in part replicates the process of succession from open ground to woodland, but the regrowth of coppiced trees is very much faster than generating a new woodland from scratch. A big storm like that of 1987 will tear trees up by the roots, creating huge amounts of ground disturbance and killing a lot of the trees: the subsequent succession to woodland is a decades-long process during which the land spends much longer as open scrub than does coppice woodland. Though lady orchids often thrive in coppice woodland, their natural home must surely be, like fly orchids, as temporary residents during the long process of the building of a wild woodland. To really work for species like this, a wild system would need to include an almost continual process of natural woodland destruction and reconstruction – orchids do not enjoy a long life as seeds in the soil – and the plants themselves would need to be able to move to find the right new homes as they came on the market.

The question of dispersal – how plants spread through the landscape – has been one of the preoccupations of this book. The importance of animals, both wild and domestic, in moving plant seeds is beyond doubt, and the European study of seeds carried by deer and wild boar has previously been mentioned. Interestingly, the 55 plant species found on the feet and fur of the animals examined in that study were not all specifically adapted to this form of transport. Some of the seeds were indeed equipped with hooks for catching onto skin and hair, like burdock *Arctium* spp. and wood avens *Geum urbanum*, but others, such as the seeds of birches *Betula* spp., were primarily adapted to being transported by wind. The seeds were also from a variety of habitats, and included species of ancient woodland such as wood millet *Milium effusum* and wood-sorrel *Oxalis acetosella*, grassland plants like autumn hawkbit *Scorzoneroides autumnalis*, and even corn spurrey *Spergula arvensis* and other cornfield weeds. At Ranscombe Farm Reserve, cattle may have carried man orchids to a new spot, and I have previously discussed how the northward migration of woodland plants through Europe after the last ice age must surely have been dependent on long-distance movements by animals. This suggests that the presence of large animals, and the ability of these animals to move widely through the landscape is essential if any rewilding project is to work for plants.

It may of course be that we will never be able to get the right sort of dispersal process in place for plants. We do not know what the wild landscape of Britain looked like when Mesolithic people first got here, but it had animals that no longer exist here or anywhere else – the massive aurochs *Bos primigenius* has been globally extinct for centuries – and it lacked our fences and motorways and cities and myriad other blocks to long-distance animal movements. Whole natural systems which once provided the necessary transport networks for wild plants may be gone, never to return, and we may never understand how they worked. At Ranscombe Farm Reserve, clustered bellflower is just one of a number of plants of old grassland which seem unable to move even a few hundred yards; some plants at Plantlife's Joan's Hill Farm nature reserve have stayed resolutely put for decades and failed to take advantage of wandering cattle. Whether they are just slow-coaches, or are waiting to hitch a lift on some process that we have not guessed and cannot replicate, the plants are giving us a strong hint that it may be hard for rewilding projects to really do the job for wild plants without at least some deliberate introductions, whether this is of seed or green hay or pot-grown plants.

Slow to colonise new sites: clustered bellflower. Photograph: R I Moyse/ Plantlife.

Perhaps the plants at Ranscombe Farm Reserve that say the most about rewilding and its role in nature conservation are those cornfield weeds which have no real home in Britain outside the man-made environment. I asked the question earlier about how, even whether, we should value these species, particularly those like common poppy and interrupted brome – perhaps even the house sparrow – that may only exist because of human activities. Well, of course we should, because they are beautiful, because they are part of our culture and often our everyday lives, and because they add life and bring joy to the places we live in. If there is an ethical, a moral, aspect to rewilding – for me at least – it is that it says that we want wild plants and animals to be able to live on their own terms and in their own ways, and not just be at the beck-and-call of us humans. But we owe that respect to all species, including those that have made their homes in the human spaces – whether these are plants of the cornfields or urban streets – and that will mean continuing to give them the space they need, and deliberately creating that space if necessary. It also means that we are allowed to tweak, redirect or interfere with rewilding if we find it is not working for wild plants and animals, because, in the end, the only judges of rewilding's success will be us.

Chapter 12 - A place for people

Ranscombe Farm Reserve is criss-crossed by footpaths: the network comes to more than 10 miles (16km) in all, and most of this is made up of public rights-of-way. As a result, it costs nothing to visit the reserve and you can come and go any day of the year and any hour of the day, though the steep slopes, and many muddy or uneven paths do mean that access remains very limited for wheelchairs or mobility scooters, despite the work that has been – and continues to be – done to try and make it easier for everyone to get around the site. With so many paths and so many people living within a short distance of the reserve, it is unsurprising that tens of thousands of visits are made every year – but probably few of the visitors think about the people whose footsteps they are treading in. Many of the footpaths at Ranscombe are hundreds of years old, perhaps following routes established by the site's earliest farmers. A pen-and-ink map from 1797, held by the British Library but available online, shows a series of tracks across the farm and through the woods that are still in place today, a couple of which – the main track through Cobham Woods, and parts of what is now the North Downs Way – were probably roads at one time.

In fact, the main track through Cobham Woods, which runs between between Strood and Cobham, is still called the Town Road, and is effectively an extension of The Street in Cobham, running eastwards towards the bridge across the Medway at Rochester. It is sometimes said that where the main road through a settlement is called simply 'The Street', this means that it was originally a Roman road. If this is the case, then it suggests that the Town Road may even date back as far as Roman times, though it's not clear to me why they would want another road parallel to Watling Street, and less than a mile to the south. Today, the Town Road though Ranscombe is a public footpath and no longer accessible to vehicles (though horses and cycles are allowed to use it), and its historic use as a route from Rochester to Cobham and beyond gives Ranscombe Farm Reserve its link with one of Britain's most famous writers, Charles Dickens.

There are, of course, a good number of places in Britain that make a claim to Charles Dickens. There are Charles Dickens museums in London and Portsmouth, and Broadstairs, on the East Kent coast, has its own Dickens Festival. But the Medway Towns have a greater claim than most. Dickens lived in Chatham as a boy, and later in life at Gad's Hill Place near Strood – less than around 2km (less than 1½ miles) north of Ranscombe – which was his home from 1856 till his death in 1870. Places in and around Medway provided settings and locations for many of Dickens' stories,

The Town Road in Cobham Woods is lined with large, old hornbeam trees. Photograph: Bob Gibbons/Plantlife.

perhaps the best known being the churchyard in the village of Cooling and Restoration House in Rochester, which both feature in Great Expectations – the first as the graveyard where Pip first meets Magwitch, and the latter as Miss Havisham's home of Satis House. A fictionalised Rochester is also very strongly the setting for The Mystery of Edwin Drood, unfinished at Dickens' death.

Dickens may have frequently taken the Town Road through Cobham Woods during the years he lived at Gad's Hill Place: he was a good friend of John Bligh, the 6th Earl of Darnley, and was a regular visitor to Cobham Hall, and was also known to frequent The Leather Bottle, a pub in the village of Cobham. However, it is clear that Dickens knew Cobham Woods long before this, as the woods make an appearance in his first book, The Pickwick Papers, published when he was in his mid-20s and living in London. In Chapter 11 of The Pickwick Papers, Mr Pickwick and his friends walk from Rochester to Cobham in order to visit a friend staying at The Leather Bottle. As the most obvious and direct route from Rochester to the village of Cobham was via the Town Road, this must have been the path along which Dickens sent his characters. In the book, Pickwick and friends find that "their way lay through a deep and shady wood, cooled by the light wind which gently rustled the thick foliage, and enlivened by the songs of the birds that perched upon the boughs" and they come out into "an open park, with an ancient hall". You can still do the same today, and also still visit The Leather Bottle on The Street in Cobham where, like Mr Pickwick, you can, if so inclined, refresh yourself with "a copious draught of ale". Dickens also gives Cobham Woods a small, 'walk-on' part in his short story, The Seven Poor Travellers, where they are mentioned by name as being part of the route the narrator takes on his way on foot from Rochester to London.

None of this turns Ranscombe into an important location in Dickens, but it does provide a tiny link from his life to ours. When we walk on the Town Road today, we know we are treading the same ground as Dickens, and we can look at the woodland around us and compare it with the way he saw the woodland when he used it in The Pickwick Papers. In this sense, we can feel a connection to Charles Dickens in the same way that archaeological finds – the stone tools, the ancient coins – link us to the Medieval, Roman and prehistoric people that once lived and worked in this landscape. This is what art, literature, music and other forms of culture can do for us: link us to other people, now and in the distant past, and give us a misty glimpse of the world seen through their eyes.

I would include the study of natural history as an aspect of our culture, too. Of course, natural history aims to be a branch of science and to give us facts rather than feelings, but the ways in which people have explored, and still explore, the natural world still gives us a view through the eyes of other people. Reading the *Iter Plantarum* of Thomas Johnson, or the reports of excursions in the 19th Century journal, The Phytologist, we get an idea of the writers' lives, the world they inhabited and the way in which they viewed the places and the plants they encountered. Their worlds have long vanished, but we can see them reflected. Better yet, we have living links to the worlds of the naturalists of the 17th, 18th and 19th Centuries in the shape of plants like meadow clary and rough mallow that those naturalists came to see at Ranscombe Farm, and which are as much a part of local cultural heritage – or even national cultural heritage, given the importance of some of the people involved – as Cobham Hall and its deer park.

Cultural heritage is not just about prominent people or those who made it into publication. Wild plants can tie us back into the lives of the ordinary people who lived side-by-side with nature in Britain for the many centuries before the industrial revolution started to split the nation into town and country. As a result, any bit of wild space can act as a kind of index of

Common poppies dominated Kitchen Field in the summer of 2020.
Photograph: R I Moyse/Plantlife.

associations and connections, and the richest sites, like Ranscombe Farm Reserve, become like fat dictionaries with many thousands of meanings to discover. You only have to look at books like Geoffrey Grigson's The Englishman's Flora, first published back in 1958, or Richard Mabey's more recent – and quite beautiful – Flora Britannica to see the richness of the cultural associations and meanings that can be found in our wild plants. Both these books are substantial works: Mabey's book runs to nearly 500 pages and covers 1,000 species, while Grigson's book is extraordinarily thorough on the various regional names given to wild plants – for example, a dozen different names are given for hazel and more than forty for scarlet pimperne.

Common poppy clearly attracted as much attention as scarlet pimpernel. It, too, has more than forty regional names given by Grigson. In Kent, for instance, you might once have heard it referred to as red-weed, red petticoats or popple. And, of course, the common poppy is one of the few wild plants that still carries a huge weight of cultural meaning in modern Britain. When I worked at Ranscombe, it was the plant that people most spoke about or telephoned about: they'd heard the site was good for poppies, and wanted to know when or where the best displays were; or they wanted to be able to photograph the poppies. This interest was only increased by the centenary of the First World War. Forty million poppies are distributed each year by the Royal British Legion around Remembrance Sunday each year, when we commemorate the dead and wounded of the 20th Century's two World Wars and of other conflicts since. Even so, it can hard to pin down precisely what the symbol of the poppy does actually mean to people. The Royal British Legion's own website says that the poppy stands for remembrance and hope for a peaceful future, but also as a show of support for the armed forces. Others – including war veterans – have complained that the poppy has been used to build support for more recent military campaigns and that the true horror of war has been forgotten. The poem which originally inspired the use of the poppy as symbol of remembrance was In Flanders Fields, written in 1915 by Canadian soldier John McCrae, and is very much an exhortation from the dead to the living to continue the fight. Given how the First World War is often viewed today – at the very least as being unnecessarily wasteful of human lives – then I'm not sure whether many people would now wholly share McCrae's attitude to that particular conflict.

A lot of the visitors who come to Ranscombe for its poppies hope to see a uninterrupted, brilliant sheet of scarlet, formed by a solid mass of thousands of plants growing together, but this kind of display is actually an uncommon event. It does happen from time to time at Ranscombe as it can do on other farmland, but seems to require a very specific set of circumstances; and perhaps it's the rarity of the sight of a whole field coloured bright red that makes it so memorable. The same is probably true of the battlefields of the First World War. We think of places like the Somme being dense with poppies growing from the churned soil, and this does appear to have been the case in places: in Where Poppies Blow, the author John Lewis-Stempel quotes a 1917 letter to The Times, which speaks of the ground flaming with poppies in areas of fierce fighting such as the Thiepval slope. But elsewhere, the trench-digging and the shelling gave rise to a variety of flowers. The same letter to The Times talks about a mix of colours – not just the red of the poppies, but the yellow of sowthistles, white of chamomile and blue of cornflowers. Something similar is seen in the painting Troops Going over the Top, First World War (Battle of the Somme) by the painter and photographer Arthur Dugmore, who had served on the Somme. Painted from memory in 1916, it shows soldiers advancing through the barbed wire while others lie dead or wounded, in a field of red and white flowers with occasional dots of blue and yellow. The look, with the exception of the cornflower blue, is very much that of summer in Ranscombe's Kitchen Field.

The idea that common poppies are linked with blood and death turns out to go back thousands of years. The bright red of their flowers must have something to do with this, though I suspect that the close association of common poppies with arable farming may be more important. It is easy to imagine how the annual cycle of cereal growing, with the death of the crop in autumn and its rebirth in spring – and the potential human disaster of crop failure – would generate its own mythology. In ancient Rome, poppies were sacred to the goddess Ceres, who was said to have taught humans how to sow and harvest crops, and who also maintained the boundaries between the worlds of the living and the dead. But the many centuries during which people and common poppies have lived side-by-side means that the plant has developed a range of other associations: it was said that poppies may give you a headache or even make you blind, and that picking one could bring on a thunder-storm. Common poppies are also said to ease pain (especially headaches, confusingly) or bring on sleep – possibly a confusion with opium poppies. The idea that common poppies can make you drowsy is even there in John McCrae's poem, which ends, "We shall not sleep, though poppies grow / In Flanders fields."

Whenever plants appear in art or literature or mythology, they tell us something about the relationship between people and the natural world. Monet's painting The Poppy Field near Argenteuil was painted in 1873 and was part of the first Impressionist exhibition – this is the one with the two pairs of figures where the whole lower-left quarter of the painting is filled with red poppies, and it is part of the movement towards painting in the open air by some of the key figures of the Impressionist movement. Later, Van Gogh was particularly attracted to poppies because of the strong colour: the Van Gogh Museum website lists seven paintings made between 1886 and 1890 which feature poppies, of which my favourite is known as Poppies and Butterflies. Painted in the south of France in the year of Van Gogh's death, 1890, the canvas is largely filled with foliage against which two white butterflies stand out alongside two richly-coloured common poppy flowers and four smaller flowers which are quite clearly prickly poppies. It is the presence of the prickly poppies which makes the painting stand out – not only does it tell us something about the plants that Van Gogh was finding in the fields around his home in Arles, but the plants are stunningly rendered with all of the distinctive shape and colour which makes the real thing so obvious when seen in the field.

Prickly poppies are so rare in Britain today that I suspect that those in Van Gogh's painting would barely register with most people. The same would probably be true in Van Gogh's native Netherlands, though a photo on the Flickr website shows they are grown in Amsterdam's botanical garden, the *Hortus Botanicus*, just a half-hour's walk from where the painting is held in the Van Gogh Museum. The point, though, still stands: so much wildlife has disappeared, and our day-to-day lives are so separated from the natural world, that our cultural connections with wild species are being lost. The rare cornfield weeds of Ranscombe – such as prickly poppy – are exciting to us precisely because of their rarity. Yes, they are lovely to look at, too, but they don't have the same meanings to us as they did to generations of our ancestors who lived alongside these plants.

This is something that really stands out in some of Shakespeare's writing. Shakespeare was writing for a mass audience, but he expected them to be familiar with a wide range of plants. In Hamlet, Ophelia falls into a brook and drowns while carrying a garland of flowers: "crow-flowers, nettles, daisies, and long purples, that liberal shepards give a grosser name, but which cold maids do Dead Men's Fingers call". It's an odd mix, but Shakespeare clearly believes that his audience will not only be familiar with 'long purples' but know the ruder names which farm-workers give them. Modern opinion seems to be that the plant he is referring to is the early-purple orchid: as discussed earlier, it is a plant which has been seen as having sexual connotations, and Shakespeare expects this to be picked up.

Shakespeare and his audience must have known their cornfield weeds, too. In Act 4 of King Lear, one of the symbols of Lear's mental illness is that he is wandering the fields wearing a head-dress made of weeds. He is described by his daughter, Cordelia: "Alack, 'tis he. Why, he was met even now / As mad as the vexed sea, singing aloud, / Crowned with rank fumiter and furrow-weeds, / With hardocks, hemlock, nettles, cuckoo-flowers, / Darnel, and all the idle weeds that grow / In our sustaining corn. – A century send forth. / Search every acre in the high-grown field, / And bring him to our eye." Hardocks are burdocks – the big, broad-leaved plants with the clinging, hooked flower-heads – which would be found, as today, alongside nettles *Urtica* spp. and hemlock *Conium maculatum* on the edges of the fields. Fumiter is fumitory – a plant of the furrows themselves, while darnel *Lolium temulentum* is an annual grass, related to perennial rye-grass, which was once a serious pest plant in arable fields but which is now virtually extinct in Britain, just turning up occasionally from accidental introductions.

In Shakespeare's Coriolanus, the audience's knowledge of cornfield weeds has to extend to understanding something of how the plants work. Corncockle, which I talked about in an earlier chapter, is one of the important cornfield weeds at Ranscombe. Now all but extinct in the wild, it survived for many centuries because its seed was so hard to remove from harvested grain: as a result, it would be accidentally sown back into the field when the next crop was planted. In the play, Coriolanus is a Roman nobleman in a time of public unrest and food shortages. He despises the common people of Rome and cannot bring himself to flatter or please them, and he even denies the hungry populace any access to supplies of corn. He believes in maintaining rule by the nobility, and that pandering to the people will cause further unrest. When he justifies himself to the other nobles, his reasoning uses the behaviour of corncockle as its central metaphor: "I say again, / In soothing them, we nourish 'gainst our senate / The cockle of rebellion, insolence, sedition, / Which we ourselves have plowed for, sowed, and scattered". In other words, Coriolanus believes that pleasing the people and giving them corn would be like sowing grain contaminated with corncockle seeds – rebellion would arise in the same way that corncockle comes up as a weed in the fields: the unintentional, unwanted but unavoidable result of the farmer's own ploughing and sowing.

Corncockle - understanding its behaviour gave Shakespeare a useful metaphor. Photograph: R I Moyse/Plantlife.

It took me one or two readings to properly understand what Shakespeare was saying here, and how the metaphor worked. And that is after I had read up on corncockle and also watched how it behaved in the field so that I understood how much it used to be dependent on repeated sowing as a contaminant of seed grain. Yet Shakespeare drops in the reference and moves on, presumably confident that his audience will need no further explanation. It makes you realise how much the natural world used to be part of the common experience of people in this country, and the depth of people's general knowledge about wild plants and other wildlife. The natural world used to play a large role in the stories we told about ourselves, but now most stories – particularly the very best ones – seem to be about what it means to be human or to be a human living amongst other humans. This is important, but it all seems unbalanced – where are all the stories about what it means to live in the world as one species amongst thousands of others?

Discussions about the place of people in nature have probably being going on since nature conservation started, perhaps much longer. The Romantic writers and artists of the late 18th Century could be seen as the start, reacting as they were to the rise of scientific inquiry and industrial progress. Today, the discussion has been given a new impetus by the idea of rewilding, and is made more urgent by our understanding that wild species are disappearing at an astonishing rate, and that it is human activities that are the cause of this. Pushing to some extent in the opposite direction is the understanding that people benefit from contact with nature and from being out in a natural environment, which is reflected in the increased expectation that wild places should provide people with opportunities for recreation. For example, national nature reserves were first established in the UK under the National Parks and Access to the Countryside Act 1949, and their purpose was the study or protection of plants, animals or geological features. This purpose was modified by the 2006 Natural Environment and Rural Communities Act, which gave national nature reserves an additional purpose of providing for "enjoyment of nature or for open-air recreation" where this did not interfere with nature conservation aims. Ranscombe Farm Reserve exists because Medway Council wanted to provide people in the densely built-up parts of Strood with somewhere green and natural and accessible right on their doorstep – and quite rightly, too – and it is for this reason that the site is shown on some maps as Ranscombe Farm Country Park.

But there is at best a tension, and at worst a real conflict, between conservation and recreation. For wild plants, there are obvious issues, such as localised soil compaction where people walk, and this can increase as people explore the land away from established paths and start to create new routes. At Ranscombe Farm, we have seen how an increase in visitor numbers leads to an increase in unofficial, so-called 'desire-line' paths, which eventually become very difficult to restrict or control, and, though the impact of these paths on plants is hard to quantify, a study of woodland parks in Warsaw in Poland found that increased public access resulted in loss of some woodland species and a reduction in the diversity of plants present. At Ranscombe Farm Reserve, more significant problems occurred where the new paths happen to coincide with the locations of rare plants. For example, where strips on the edges of arable fields were deliberately cultivated to encourage germination of rare plants such as ground-pine and broad-leaved cudweed, it was often the case that these worked-over strips were quickly taken up as new walking routes by some dog-walkers. The result was that the soil quickly became compacted and plant germination was reduced.

The case is even clearer with birds, where more research has been undertaken. Although Ranscombe Farm Reserve is managed primarily for its rare plants, a lot of consideration is also given to the animals that depend on those plants. This is especially important if you want to be able to show how creating a plant-rich environment has conservation benefits beyond the plants themselves. However, there is a lot of evidence that the presence of people, and especially people with dogs, can cause problems to birds year-round, disrupting both breeding in the summer and feeding in the winter. Most of the scientific work on bird disturbance has been done on heathlands or coastal sites, where the presence of particular breeding or wintering birds are the reasons why some sites are legally protected (the Medway Estuary, for instance). But there is no reason why similar impacts may not be seen on farmland or in woodland: for instance, a study in the USA, looking at recreational trails in forest habitats, found that there were greater densities of birds in places with fewer trails. Birds may be put off from nesting where they are repeatedly disturbed early in the breeding season; and, where a bird has already made a nest, any disturbance which causes the adult to leave can result in eggs being taken by predators. In winter, when birds need to feed well to simply stay alive, repeated disturbance may reduce the time they can spend feeding and lower their chance of survival. Because birds recognise dogs as predators, they often react more to a person with a dog than to someone without, and react even more to a dog which is running free: another American study found that grassland birds were particularly sensitive to lone dogs.

Broad-leaved cudweed growing on a footpath. Though repeatedly crushed by walkers' boots, it keeps flowering. Photograph: R I Moyse/Plantlife.

The story is not entirely negative, though. Trampling and soil compaction may be bad for plants, but the bare ground of footpaths is an important nesting habitat for a good number of different bee species. And not all plants are worried about being trampled over. Broad-leaved cudweed seems to be very resistant to this sort of mistreatment, and even repeatedly flattened plants seem able to flower and set seed. This may even be a useful adaptation for this very rare cornfield weed, as its seeds seem to have a knack for getting hold of walkers' feet: a regular visitor to Ranscombe Farm once found a broad-leaved cudweed plant growing between the bricks on his driveway, some two miles from the reserve, the seed presumably having hitched a lift on his boots. In a countryside without free-roaming wild animals, human walkers – with or without dogs – may provide an important form of free transport for the seeds of wild plants. In his 1961 book on Weeds and Aliens, the botanist Sir Edward Salisbury famously describes how he raised over 300 individual plants of more than 20 species from seeds shaken out of his trouser turn-ups. Although turn-ups on trousers have fallen out of fashion these days, Salisbury also suggests –

from his own observations – that as many as 200 seeds might be carried by a pair of boots, and notes that another researcher found the seeds of 43 different plants in the mud scraped from boots and shoes.

One of the options that has been proposed for the future conservation of wildlife in Britain is a greater emphasis on so-called 'land sparing', where nature conservation is concentrated in fewer, much larger areas, leaving the rest of the countryside to be intensively managed for agriculture. Studies suggest that this would give better results for wildlife than trying to generally improve the wider landscape for biodiversity. The trouble with this, for me, is that it seems one inevitable result of land sparing would be most people finding themselves further away from places rich in wildlife: nature would become something you travelled to, not something on your doorstep, and even less a part of people's day-to-day lives. It is the same thing that worries me when I think about the way in which we affect wildlife through our recreational use of wild places and nature reserves. We must find our way towards a shared understanding of how to use our wild places for fun – for walking the dog, for playing with the children, for just wandering aimlessly – without crushing all the plants or scaring off the birds. Otherwise, we could end up with a future where there are places where anyone can go but which are dull and wildlife-free, and other places full of wild species where only naturalists are allowed to go.

The art historian, Dr James Fox, argues in his recent television series, Nature and Us: A History through Art, that the development of farming marked a major change in our relationship with the natural world, at least in Europe and the Middle East. His view is that when people still lived as hunter-gatherers, they saw themselves as part of nature; once farming became established, the natural world became something to be controlled, and something separate to human society. He cites the differences between early European cave paintings – with their naturalistic portraits of wild horses, cattle, bears and lions, and apparent reverence for them – with the huge bas-relief sculpture of a lion hunt from the palace of the ancient Assyrian King Ashurbanipal (currently in the British Museum). The cave paintings rarely include depictions of people, but instead focus on the power and fearfulness of the animals depicted, and James Fox draws parallels between the paintings and more modern art from native cultures in Africa and North America that see animals as spiritual beings with their own souls and even magical powers. In contrast, the lion hunt images are all about the people – particularly the king himself – and their power over the animals, which have been deliberately captured and released so that they can be hunted and slain by Ashurbanipal and his entourage. At the time the sculpture was made, around 2,500 years ago, Assyria would have been an agricultural society for many centuries.

The idea that nature is separate to us humans, and is there to serve us, and to be directed and controlled, colours our entire relationship with the other species on this planet. It has certainly shaped much of conservation and continues to do so, even people's intentions are good. Even the idea of 'ecosystem services' – that the natural world provides us with useful life-support systems, like reducing flooding or cleaning the air – could be seen in this way, though the concept was originally promoted as a way of demonstrating the importance of nature conservation to society. When you look at figures from the Office for National Statistics (ONS), everything about ecosystem services is reduced to a monetary value: for example, the recreational value of woodland in England is estimated by the ONS at just short of half-a-billion pounds a year on average. This makes us consumers of the natural environment, particularly when you see that monetary values are also given for the work that woodlands do in removing carbon dioxide and other pollution from the atmosphere. This kind of 'trade' arrangement – we make a mess and woodlands clear it up for us – devalues and undermines the natural world. What would happen if we stopped polluting: does woodland wildlife suddenly become less important to us and less worthy of conservation?

This does not mean that the benefits we derive from nature, those ecosystem services, are not real, because they are very real indeed. Likewise, ideas about 'land sparing' to get better results for our conservation efforts are hugely valuable – you cannot and should not deny the science behind either approach. But both have the potential to increase the separation between nature and ourselves, at a time when it has never been more important to close that gap.

I am not sure, though, that I entirely agree with James Fox that the development of farming marked the point when we saw ourselves as separate from nature. Yes, it was undoubtedly an important step on that journey, but I suspect that the change was more gradual, with our disconnection from the wild world increasing as our power over it has increased. For most of the history of farming, we did not have the systems, the chemicals or the machinery to completely subdue the wild. We could not keep the weeds from the cornfields or the birds from the crops. People had to live side-by-side with other species. The folk names, the traditions and the superstitions associated with so many wild plants and animals show how this constant rubbing of shoulders meant that we saw wild species as having their own powers, abilities and knowledge. One of the great

Dulse *Palmaria palmata* - a seaweed used in magic in the Medieval British stories of the Mabinogion.

treasures of British literature is the collection of Medieval Welsh stories known as The Mabinogion, which were probably first written down between the late 11th Century and the early 14th Century but undoubtedly go back much earlier as part of an oral tradition of story-telling. Even though the stories are both set and written thousands of years after the arrival of farming in Britain, the world of The Mabinogion is one where wild animals, and even to some extent wild plants, may be as active and important as people. Particularly striking is the group of the oldest animals that appear, and help the heroes, in the story of how King Arthur and his nobles help Culhwch to find and marry Olwen, daughter of the chief giant, Ysbaddaden. These animals include the Blackbird of Cilgwri, who has lived there long enough to have worn away a whole anvil with its beak; the Stag of Rhedynfre, who has watched an oak grow from a sapling to a mature tree then die and rot away; and the Salmon of Llyn Lliw, a mighty beast who has survived having fifty tridents removed from its back and who carries two of Arthur's men up-river. Even wild plants have magical qualities in some of these old tales. In the fourth of the linked stories known as the Four Branches of the Mabinogi, the magician Gwydion turns seaweed – dulse *Palmaria palmata* and wrack *Fucus* spp. are specifically mentioned – into fine leather so that he can pose as a shoemaker; later in the same tale, a living woman is conjured from the flowers of oak, broom and meadowsweet.

The tales in The Mabinogion were not stories created for children, but as entertainments for adults living in a world that was shared with wild species that they could not control or fully understand, but with which they felt some affinity. Certainly, the more time I have spent amongst wild plants and animals, the less – and less important – the differences seem between them and us. I mentioned in a previous chapter how watching a red-tailed mason bee at work led me to wonder whether it felt a sense of satisfaction in its work. Ascribing emotion to a tiny insect is something that grates uncomfortably on my sense of scientific detachment, but when you catch bees to examine them, it is astonishing to see the wide range of their responses. Some will sit quietly in the glass tube. Others roam around before stopping and giving themselves a good clean. Yet others – and the mightily-jawed mason bees are amongst them – will frantically try to chew their way to freedom. Something is going on inside these little animals, and the proper scientific response is to suggest that they are acting on instinct, whereas a human in a similar predicament would be acting according to feelings or reason. Yet, modern thinking seems to be that human free will is an illusion, and all our rationalising about why we act in a certain way is no more than a construction to cover for decisions that are instinctive or at best unconscious. If this is true, then it is hard to say why we should ascribe inner lives to ourselves but not to most other animals.

I won't argue for plants being thinking or feeling beings, even though more and more information is coming to light about the ways in which they are able to communicate with each other, and the discovery of electrical impulses within and between plant cells. Yet they still have their struggles and their challenges and their ways of responding to the world around them, even if this is completely unknown to them and is only apparent to a human observer. For this reason, I'm more than happy to greet a familiar plant as if it is an old friend, and I think we could do with more stories where people and plants and animals meet on a more equal footing. Not magical tales with talking blackbirds and ancient salmon, and not the kinds of Hollywood epics that allow us to cast ourselves as the heroes that save the day, but soap operas which have a big, diverse cast and are full of incidents, cliff-hangers, unexpected twists and character back-stories. Of course we need plenty of cold, hard science to tell us how we can save wildlife, but only our stories will tell us why we should bother.

Maybe Ranscombe Farm Reserve is important to me because it holds so many stories. The land itself, the fields, woods, the tracks and the boundaries tell us about thousands of years of human history. Evidence of lost crofts, the well in Kitchen Field, the tithe award that gave Bishop's

Wood its name – even perhaps a 13th Century mill on Mill Hil – link us back to the time when the fantastical stories in The Mabinogion were being written down. Just by putting our feet on the ground, we tread in the footsteps of everyone from Mesolithic hunter-gatherers to Charles Dickens. We can, without too much trouble, meet the direct descendants of the plants that met the person who sent Darwin to explore the world on HMS Beagle. There are plants in the ancient woodland whose stubborn tendency to stay put might help us understand how they got here after the last Ice Age, and what that might mean for modern-day attempts at rewilding. Most of all, perhaps, there are the cornfield weeds which tell a story of 10,000 years of co-existence – plants that have sown themselves into our shared cultural space, whether this is in the art of the Impressionist painters, the works of William Shakespeare or our still deeply-held remembrance of the horrors of the First World War.

"Who Do You Think You Are?" is the question asked by the popular BBC TV programme that traces the genealogy of various celebrities. We like to know our own back-stories, our own links to the past and our place in wider history. Plants and other wildlife can tell us this. The wide diversity of colourful cornfield weeds at Ranscombe Farm hitched a lift with our own predecessors as they travelled from the Middle East, bringing farming with them, and link our culture with theirs. The common poppies which we so revere as symbols of remembrance were sacred to Ceres in ancient Rome; to the ancient Greeks, they were the flowers of Aphrodite; in Ashurbanipal's Assyria, they were the 'daughters of the field'. Even the medieval Welsh tales of The Mabinogion include international themes: in Sioned Davies's translation, she points out that the story of the oldest animals has parallels in Indian and Persian literature, and indeed the whole story of Culhwch and Olwen reflects an international folk-tale motif. The stories and meanings and mythologies associated with nature are shared half-way around the world or even further. The return of ravens to Ranscombe is exciting not just because it represents a successful conservation story, but because ravens live right around the northern hemisphere and have been important in the mythology of every culture. In Judaeo-Christian culture, a raven was the first bird released by Noah from the ark. The raven is central to creation myths of native people of North America, and is also a trickster there as it is in Siberian mythology; while in Norse mythology, two ravens were the associates of the chief god, Odin. An important and supernatural figure in one of the stories of The Mabinogion is Bendigeidfran, or Brân the Blessed, where Brân can be translated as 'raven'.

While I was out one day at Ranscombe, I met an East Asian lady who had been foraging for plants for use in traditional Chinese medicine.

Amongst the material she had collected were leaves of greater plantain *Plantago major*. This is a very common plant, with broad, round, strongly-ribbed leaves, and which can withstand heavy trampling so that it often grows on the bare mud of footpaths. The medicinal use of greater plantain was new to me, but a little bit of online research shows that plantains are widely used in traditional Chinese medicine for a range of conditions. This turns out to be another shared tradition, because greater plantain was a sacred healing herb to the Anglo-Saxons, and is still regarded as being useful in western herbal medicine. If there is one good argument for the place of people in nature – and for a countryside which is inclusive and accessible to all – it is this: that our shared relations with the natural world can actually bring us closer together as people.

Acknowledgements

I would like to say a big 'thank you' to Joe Costley, Dave May, Helen Morley and Ben Sweeney, who all provided valuable information which has informed the writing of this book, and to Plantlife for allowing use of many of the photos herein. In particular, Dave's work on the site's archaeology and history has been vital to telling the story of the reserve. Thank you, too, to Dr Simon Springate and to Ian Tittley for proof-reading and comments, and especially to my partner, Kathy Friend, who gave so much encouragement and who read and commented on early drafts of many chapters.

But I would also like to give credit to all those people without whom Ranscombe Farm Reserve would not have happened – or at least would not have become the extraordinary place it is – and who did not get a mention in the main text. At Plantlife, this includes the trustees, the three chief executives I worked under, and all the staff and volunteers in the fund-raising, finance, communications, education, conservation and office management teams. Plantlife is not a large organisation, and I don't think there was a single person there who didn't contribute in some way, from raising the cash, to supporting events and outreach, advising on plant ecology, or keeping our laptops running. Big thanks are especially due to Tim and Jacqui who did such great work developing and running outreach and education at Ranscombe, an important aspect of the reserve's work which I have not touched on in this book. And it is essential to acknowledge the many hours of hard work contributed by the scores of volunteers who have worked at Ranscombe over the years, carrying out practical work, managing livestock, monitoring wildlife and picking litter – I'm sorry that I am unable to name you all here.

I must also thank the elected members and officers at Medway Council who have supported Ranscombe Farm Reserve, and bodies including the Kent Downs AONB, Valley of Visions Project, Cuxton Parish Council, West Kent Downs Countryside Trust, Cuxton Countryside Group, the National Trust and Kent Wildlife Trust, all of whom contributed support, funding, advice or person-power.

References

I have tried here to give a list of all the sources I consulted when putting this book together. As it is quite a long list and includes a mix of popular writing, web pages and scientific papers, I have broken the references down by broad subject area, to make it easier for the browser to find what they might be interested in. I have only put internet addresses where the source is a specific web page, but all the scientific papers can also be found online (though in many cases only an abstract of the full paper is available without further payment).

Plants, plant lore and plant people

Andronova, E.V., Kovaleva, A. A., Evdokimova, E. E., Nazarov, V. V. & Semeonov, A. V. 2020. Fruitage and seed viability of *Orchis purpurea* (Orchidaceae). *The International Journal of Plant Reproductive Biology* vol. 12, pp56-66.

Badr, A., Rabey, H.E., Effgen, S., Ibrahim, H.H., Pozzi, C., Rohde, W. and Salamini, F. 2000. On the origin and domestication history of barley (*Hordeum vulgare*). *Molecular biology and evolution* vol. 17, pp499-510.

Balfour, N. J., & Ratnieks, F. L. 2022. The disproportionate value of 'weeds' to pollinators and biodiversity. *Journal of Applied Ecology*.

Barke, B. H., Karbstein, K., Daubert, M., & Hörandl, E. 2020. The relation of meiotic behaviour to hybridity, polyploidy and apomixis in the *Ranunculus auricomus* complex (Ranunculaceae). *BMC plant biology* vol. 20, pp1-14.

Brunet, J., & von Oheimb, G. 1998. Colonization of secondary woodlands by *Anemone nemorosa*. *Nordic Journal of Botany* vol. 18, pp369-377.

Byfield, A. 2010. Is nature conservation working for plants? *British Wildlife* vol. 22, pp31-40.

Cooper, M.R. & Johnson, A.W. 1984. *Poisonous Plants in Britain and their effects on animals and man*. Her Majesty's Stationery Office.

Costley, J. 2015. Meadow management increased plant species diversity in a species-poor, neutral grassland in Herefordshire, UK. *Conservation Evidence* vol. 12, pp40-42.

Gilmour, J.S.L. (ed.). 1972 *Thomas Johnson - Botanical Journeys in Kent & Hampstead. A facsimile reprint with Introduction and Translation of his* Iter Plantarum *1629*, Descriptio Intineris Plantarum *1632*. The Hunt Botanical Library.

Grant, W. F., & Small, E. 1996. The origin of the *Lotus corniculatus* (Fabaceae) complex: a synthesis of diverse evidence. *Canadian Journal of Botany* vol. 74, pp975-989.

Grigson, G. 1958. *The Englishman's Flora* (1975 edition). Paladin.

Harold, B. 1998. *Potentilla erecta/P. reptans/P.anglica* and hybrids. In Rich, T.G.C. & Jermy, A.C. Plant Crib. BSBI.

Harrower, C.A., Rorke, S.L. & Roy, H.E. 2021. *England Biodiversity Indicators 2021: Pressure from invasive species – technical background documen*t. Department for Environment, Food & Rural Affairs.

Hegland, S. J., Van Leeuwen, M., & Oostermeijer, J. G. B. 2001. Population structure of *Salvia pratensis* in relation to vegetation and management of Dutch dry floodplain grasslands. *Journal of Applied Ecology* vol. 38, pp1277-1289.

Heinken, T., & Raudnitschka, D. 2002. Do wild ungulates contribute to the dispersal of vascular plants in central European forests by epizoochory? A case study in NE Germany. *Forstwissenschaftliches Centralblatt Vereinigt Mit Tharandter Forstliches Jahrbuch* vol. 121, pp179-194.

Hörandl, E. 1998. Species concepts in agamic complexes: Applications in the *Ranunculus auricomus* complex and general perspectives. *Folia Geobotanica* vol. 33, pp335-348.

Irvine, A. 1844. Short Account of an Excursion to Cobham, Kent. *The Phytologist: Popular Botanical Miscellany*, vol. 1.

Jacquemyn, H., Brys, R., & Hutchings, M.J. 2011. Biological flora of the British Isles: *Orchis anthropophora* (L.) All. (*Aceras anthropophorum* (L.) W.T. Aiton). *Journal of Ecology* vol. 99, pp1551–1565.

Jersáková, J. & Malinová, T. 2007. Spatial aspects of seed dispersal and seedling recruitment in orchids. *New Phytologist* vol. 176, pp237-241.

Jury, S. L. 1977. The herbarium of J. E. Lousley. *Watsonia* vol. 11, p312.

Kadereit, J. W. 1990. Some suggestions on the geographical origin of the central, west and north European synanthropic species of *Papaver* L. *Botanical Journal of the Linnean Society* vol 103, pp221–231.

Kirby, K. 2020. *Woodland Flowers – colourful past, uncertain future*. Bloomsbury Wildlife.

Kitchener, G. 2018. Kent Rare Plant Register Draft species accounts: M. Kent Botanical Recording Group.

Kitchener, G. 2021. Kent Rare Plant Register Draft species accounts: S (first part: Sa-Sera). Kent Botanical Recording Group.

Leipold, M., Tausch, S., Poschlod, P., & Reisch, C. 2017 Species distribution modeling and molecular markers suggest longitudinal range shifts and cryptic northern refugia of the typical calcareous grassland species *Hippocrepis comosa* (horseshoe vetch). *Ecology and Evolution* vol. 7, pp1919-1935.

Lewis, D. 2012. *Thomas Johnson (1600 - 1644) – The father of British Field Botany*. Yorkshire Philosophical Society. https://www.ypsyork.org/resources/yorkshire-scientists-and-innovators/thomas-johnson/.

Lodwick, L. 2014. *The corncockle kerfuffle – the archaeobotanical evidence*. https://theplantremains.wordpress.com/.

Lousely, J.E. 1950. *Wildflowers of Chalk and Limestone*. Collins.

Mabey, R. 1996. *Flora Britannica*. Sinclair-Stevenson.

Mabey, R. 2021. Scribbling in the margins. *British Wildlife* vol. 33, p45.

Mandáková, T., Marhold, K., & Lysak, M. A. 2014. The widespread crucifer species *Cardamine flexuosa* is an allotetraploid with a conserved subgenomic structure. *New Phytologist* vol. 201, pp982-992.

Marren, P. 2006. The last stand of the wild flower. The Independent. https://www.independent.co.uk/climate-change/news/the-last-stand-of-the-wild-flower-414991.html.

Marren, P. 2019. Comment: How important is native status? Conservation and alien plants. *British Wildlife* vol. 30, pp427-432.

Metherell, C. & Rumsey, F.J. 2018. *Eyebrights (*Euphrasia*) of the UK and Ireland: BSBI Handbook No. 18*. BSBI.

Pearman, D. 2019. The impact of native and alien 'thugs' in semi-natural habitats. *BSBI News* no. 142, pp6-11.

Peterken, G. 2000. Identifying ancient woodland using vascular plant indicators. *British Wildlife* vol. 11, pp153-158.

Peterken, G. 2013. *Meadows*. British Wildlife Publishing.

Plukenet, L. 1696. *Almagestum Botanicum sive Phytographiae Pluknetianae Onomasticon*.

Porter, M. & Foley, M. 2017. *Violas of Britain and Ireland: BSBI Handbook No. 17*. BSBI.

Preston, C.D., Pearman, D.A. & Hall, A.R. 2004. Archaeophytes in Britain. *Botanical Journal of the Linnean Society* vol. 145, pp257–294.

Rich, T. C. G., & Lockton, A. J. 2002. *Bromus interruptus* (Hack.) Druce (Poaceae) - an extinct English endemic. *Watsonia* vol. 24, pp69-80.

Robinson, M. 2011) The paleoecology of alluvial hay meadows in the Upper Thames valley. *Fritillary* vol 5, pp47-57.

Rogers, S. 1969. Studies on British poppies I. Some observations on the reproductive biology of the British species of *Papaver*. *Watsonia* vol. 7, pp55-63.

Rose, F. 1999. Indicators of ancient woodland – the use of vascular plants in evaluating ancient woods for nature conservation. *British Wildlife* vol. 10, pp241-251.

Rose, F. 2006. The *Wildflower Key* (revised and expanded by Clare O'Reilly). Frederick Warne.

Salisbury, E. 1964. *Weeds and Aliens*, 2nd edition. Collins.

Scholz, H. 2008. Some comments on the genus *Bromus* (Poaceae) and three new species. *Willdenowia* vol. 38, pp411-422.

Shellswell, C.H. and Waller, M. (eds.) 2020. *Looking after interrupted brome* Bromus interruptus*: Ecology and Conservation Portfolio*. Plantlife.

Stace, C. 2019. *New Flora of the British Isles*, 4th edition. C & M Floristics.

Stace, C. A., Preston, C.D., & Pearman, D.A. 2015. *Hybrid Flora of the British Isles*. Botanical Society of Britain and Ireland.

Stevens, C.J., Wilson, J., & McAllister, H.A. 2012. *Biological Flora of the British Isles*: Campanula rotundifolia. *Journal of Ecology* vol. 100, pp821–839.

Strange, P. 2015. *The Opium Fields of England*. Philip Strange Science and Nature Writing. https://philipstrange.wordpress.com/2015/09/03/the-opium-fields-of-england/.

Stroh, P.A., Leach S.J., August, T.A., Walker, K.J., Pearman, D.A., Rumsey, F.J., Harrower, C.A., Fay, M.F., Martin, J.P., Pankhurst, T., Preston, C.D. & Taylor, I. 2014. *A Vascular Plant Red List for England*. BSBI.

Symons, J. 1798. Synopsis *Plantarum insulis Britannicis indigenarum*.

Unnamed author. 1839. Details of an Excursion from Woolwich, Kent to Cobham and Cuxton. *Proceedings of the Botanical Society of London* vol. 1, pp82-87.

Unnamed author. 2012. *Our vanishing flora – how wild flowers are disappearing across Britain*. Plantlife.

Unnamed author. *Darwin Correspondence Project – John Stevens Henslow*. University of Cambridge. https://www.darwinproject.ac.uk/john-stevens-henslow.

Unnamed author. *Hyacinthoides non-scripta* (L.) Chouard. Royal Botanic Gardens, Kew – Plants of the World Online.

Unnamed author. *Plukenet, Leonard (1642-1706)*. Global Plants Database. JStor. https://plants.jstor.org/stable/10.5555/al.ap.person.bm000047919.

Unnamed author. *The Herbal or General History of Plants by John Gerard and Thomas Johnson*. 2018. The Ex-Classics Project. https://exclassics.com/herbal/herball.txt.

Völler, E., Auge, H., Bossdorf, O., & Prati, D. 2013. Land use causes genetic differentiation of life-history traits in *Bromus hordeaceus*. *Global Change Biology* vol. 19, pp892-899.

Insects, birds and mammals

Balmer, D.E., Gillings, S., Caffrey, B.J., Swann, R.L., Downie, I.S. & Fuller, R.J. 2013. *Bird Atlas 2007-11: the breeding and wintering birds of Britain and Ireland*. BTO Books.

Benton, T. 2012. *Grasshoppers and Crickets*. Collins.

Brook, J. & Brook, G. 2009. *Dragonflies of Kent* (revised edition). Kent Field Club.

Clements, R., Orchard, M., McCanch, N. & Wood, S. 2015 *Kent Breeding Bird Atlas 2008-13*. Kent Ornithological Society.

Corbet, P. & Brooks, S. 2008. *Dragonflies*. Collins.

Crawley, D., Coomber, F., Kubasiewicz, L., Harrower, C., Evans, P., Waggitt, J., Smith, B. & Mathew, F. 2020. *Atlas of the Mammals of Great Britain and Northern Ireland*. Pelagic Publishing.

Easterbrook, M. 2010. *Butterflies of Britain & Ireland – a field and site guide*. A & C Black.

Eaton, M., Aebischer, N., Brown, A., Hearn, R., Lock, L., Musgrove, A., Noble, D., Stroud, D. & Gregory, R. 2015. Birds of Conservation Concern 4: the population status of birds in the UK, Channel Islands and Isle of Man. *British Birds* vol. 108, pp708-746.

Falk, S. & Lewington, R. 2015. *Field Guide to the Bees of Great Britain and Ireland*. Bloomsbury.

Heinken, T., Hanspach, H., Raudnitschka, D., & Schaumann, F. 2002. Dispersal of vascular plants by four species of wild mammals in a deciduous forest in NE Germany. *Phytocoenologia* vol.32, pp627-644.

Johnson, H. 2017. *Conservation Strategy for West-European Hedgehog (*Erinaceus europaeus*) in the United Kingdom (2015-2025), version 2*. People's Trust for Endangered Species & British Hedgehog Preservation Society.

Judge, J., Wilson, G. J., Macarthur, R., McDonald, R. A., & Delahay, R. J. 2017. Abundance of badgers (*Meles meles*) in England and Wales. *Scientific Reports* vol. 7, pp1-8.

Kurek, P., Kapusta, P., & Holeksa, J. 2014. Burrowing by badgers (*Meles meles*) and foxes (*Vulpes vulpes*) changes soil conditions and vegetation in a European temperate forest. *Ecological Research* vol. 29, pp1-11.

Macdonald, D.W., Doncaster, P., Newdick, M., Hofer, H., Mathews, F. & Johnson, P.J. 2015. Foxes in the Landscape: ecology and sociality. In *Wildlife Conservation on Farmland, volume 2 – Conflict in the Countryside*. Edited by D. W. Macdonald & R. E. Feber. Oxford University Press.

Macdonald, D.W., Newman, C. & Buesching, C.D. 2015. Badgers in the rural landscape – conservation paragon or farmland pariah? In *Wildlife Conservation on Farmland, volume 2 – Conflict in the Countryside*. Edited by D. W. Macdonald & R. E. Feber. Oxford University Press.

Meek, M. & Bullion, S. 2012. Can the harvest mouse survive in a modern arable landscape? A Suffolk case study. *British Wildlife* vol. 23, pp419-423.

O'Connell, E., Inns, T. & Walsh, B. 2015. *Health effects of exposure to setae of oak processionary moth larvae*. Systematic review. Public Health England.

Panter, C.J. & Dolman, P.M. 2012. Mammalian herbivores as potential seed dispersal vectors in ancient woodland fragments. *Wildlife Biology* vol. 18, pp292-303.

Perrow, M. & Jowitt, A. 1995.What Future for the harvest mouse? *British Wildlife* vol. 6, pp356-365.

Powney, G.D., Carvell, C., Edwards, M., Morris, R. K., Roy, H. E., Woodcock, B. A., & Isaac, N. J. 2019. Widespread losses of pollinating insects in Britain. *Nature Communications* vol. 10, article no. 1018.

Ravinet, M., Elgvin, T. O., Trier, C., Aliabadian, M., Gavrilov, A., & Sætre, G. P. 2018. Signatures of human-commensalism in the house sparrow genome. *Proceedings of the Royal Society B* vol. 285: 20181246. https://doi.org/10.1098/rspb.2018.1246.

Steffan-Dewenter, I. & Tscharntke, T. 2001. Succession of bee communities on fallows. *Ecography* vol. 24, pp83-93.

Sterling, P., Parsons, M., & Lewington, R. 2012. *Field Guide to the Micro-moths of Great Britain and Ireland.* British Wildlife Publishing.

Stubbs, A.E. & Falk, S.J. 2002. *British Hoverflies – an illustrated identification guide.* British Entomological and Natural History Society.

Taylor, D.W., Davenport, D.L. & Flegg, J.J.M. 1984. *The Birds of Kent*, 2nd edition. Meresborough Books.

Thomas, J. & Lewington, R. 1991. *The Butterflies of Britain & Ireland.* Dorling Kindersley.

Waring, P., Townsend, M. & Lewington, R. 2003. *Field Guide to the Moths of Great Britain and Ireland.* British Wildlife Publishing.

Young, J.S., Ryan, H., Thompson, S., Newcombe, M. & Puckett, J. 2015. *Mammals of Kent.* Kent Mammal Group, Kent Bat Group, East Kent badger Group and Kent Field Club.

Places, history, and ancient stories

Barton, N. & Roberts, A. 2004. The Mesolithic Period in England: current perspectives and new research. In Savile, A. (ed) *Mesolithic Scotland and its Neighbours. The Early Holocene Prehistory of Scotland, its British and Irish context, and some Northern European Perspectives*, pp339-358. Society of Antiquaries of Scotland.

Bilow, A. 2009. *Royal African Company*. Blackpast. https://www.blackpast.org/global-african-history/groups-organizations-global-african-history/royal-african-company-2/.

Champion, I.K. & Floodgate, E.E. (eds). 1977. *Natural History in the Rochester Area*. Meresborough Books.

Church, D. 1976. *Cuxton – a Kentish Village*. Arthur J. Cassell.

Conneller, C., & Higham, T. 2015. The early Mesolithic colonisation of Britain: preliminary results. *No stone unturned. Papers in honour of Roger Jacobi*, pp157-66.

Cox, M. & Attenborough, D. 1975. *David Attenborough's Fabulous Animals*. BBC.

Cruickshanks, E. 1983. Williamson, Joseph (1633-1701), of Whitehall and Cobham Hall, Kent. In *The History of Parliament: the House of Commons 1660-1690*, edited by B.D. Henning. https://d6.historyofparliamentonline.org/volume/1660-1690/member/williamson-joseph-1633-1701.

Hayden, C. & Stafford, E. 2006. *The Prehistoric Landscape at White Horse Stone, Aylesford, Kent*. CTRL Integrated Site Report Series. http://archaeologydataservice.ac.uk/archives/view/ctrl/whsisr/downloads.cfm. 10.5284/1008829.

Hendry, L. Undated. *Human Evolution - First Britons*. Natural History Museum website.

Lewis-Stempel, J. 2016. *Where Poppies Blow – the British Soldier, Nature, the Great War*. Weidenfeld & Nicolson.

Moore, D. 2019. *Cement Plants and Kilns in Britain and Ireland*. www.cementkilns.co.uk.

Pope, M., Wells, C., Scott, B., Maxted, A., Haycon, N., Farr, L., Branch, N. & Blinkhorn, E. 2019. *Resource Assessment and Research Agenda for the Upper Palaeolithic and Mesolithic periods: The Upper Palaeolithic and Mesolithic Periods*. South East Research Framework.

Pryor, F. 2011. *Overview: From Neolithic to Bronze Age, 8000 - 800 BC*. BBC website (archived pages).

Scott Robertson, W.A. 1877. Six Wills Relating to Cobham Hall. *Archaeologia Cantiana* vol. 11 pp274-293.

Sheridan, A. 2010. The Neolithization of Britain and Ireland: The 'Big Picture'. In Finlayson, B. & Warren, G. (eds) *Landscapes in Transition*. Oxbow Books.

Smalley, R. 2016. *The Cobham and cricket connection*. Shorne Woods Archaeology Group. http://shornewoodsarchaeology.co.uk/cobham-cricket-connection.

Smith, O., Momber, G., Bates, R., Garwood, P., Fitch, S., Pallen, M., Gaffney, V. & Allaby, R.G. 2015. Sedimentary DNA from a submerged site reveals wheat in the British Isles 8000 years ago. *Science* vol. 347, pp998-1001.

Tuson, D. 2007. *The Kent Downs*. Tempus Publishing.

Unnamed author. *National Heritage List for England – Cobham Hall. Official list entry*. https://historicengland.org.uk/listing/the-list/list-entry/1000182.

Unnamed author. 2007. *The Mabinogion – Translated with an Introduction and Notes by Sioned Davies*. Oxford University Press.

Waller, J.G. 1877. The Lords of Cobham, their Monuments, and the Church. *Archaeologia Cantiana* vol. 11, pp49-112.

White, G. 1949. *The Natural History of Selborne, edited with an introduction by R.M. Lockley*. J.M. Dent & Sons.

Nature conservation and land management

Carver, S. & Convery, I. 2021. Rewilding: time to get down off the fence? *British Wildlife* vol. 32, pp246-255.

Carver, S. 2016. Rewilding… conservation and conflict. *ECOS* vol. 37 pp2-10.

Finch, T., Day, B.H., Massimino, D., Redhead, J.W., Field, R.H., Balmford, A., Green, R.E. & Peach, W.J. 2021. Evaluating spatially explicit sharing-sparing scenarios for multiple environmental outcomes. *Journal of Applied Ecology* vol. 58, pp655-666.

Fuller, R. & Gilroy, J. 2021. Rewilding and intervention: complementary philosophies for nature conservation in Britain. *British Wildlife* vol. 32, pp258-267.

Miller, S. G., Knight, R. L., & Miller, C. K. 2001. Wildlife Responses to Pedestrians and Dogs. *Wildlife Society Bulletin* vol. 29, pp124–132.

Perino, A., Pereira, H.M., Navarro, L.M., Fernández, N., Bullock, J.M., Ceaușu, S., Cortés-Avizanda, A., van Klink, R., Kuemmerle, T., Lomba, A. & Pe'er, G. 2019. Rewilding complex ecosystems. *Science* vol. 364, article no.6438.

Sandom, C. 2016. Rewilding – implications for nature conservation. *ECOS* vol. 37, pp24-28.

Showler, D.A., Stewart, G.B., Sutherland, W.J. & Pullin, A.S. 2010. What is the impact of public access on the breeding success of ground-nesting and cliff-nesting birds? *CEE review 05-010 (SR16)*. Collaboration for Environmental Evidence. www.environmentalevidence.org/SR16.html.

Sikorski, P., Szumacher, I., Sikorska, D., Kozak, M. & Wierzba, M. 2013. Effects of visitor pressure on understory vegetation in Warsaw forested parks (Poland). *Environmental Monitoring and Assessment* vol. 185, pp5823–5836.

Soga, M., Yamaura, Y., Koike, S. & Gaston, K.J., 2014. Land sharing vs land sparing: does the compact city reconcile urban development and biodiversity conservation? *Journal of Applied Ecology* vol. 51, pp1378-1386.

Taylor, K., Anderson, P., Taylor, R., Longden, K. & Fisher, P. 2005. *English Nature Research Reports Number 649 – Dogs, access and nature conservation*. English Nature.

Thompson, B. 2015. Recreational Trails Reduce the Density of Ground-Dwelling Birds in Protected Areas. *Environmental Management* vol. 55, pp1181-1190.

Index of species mentioned in the text

Entries in bold type are species with well attested recent records at Ranscombe Farm Reserve.

Acanthis cabaret **See redpoll, lesser**
Accipter nisus **See sparrowhawk**
Acer campestre **See maple, field**
Achillea millefolium **See yarrow**
Adder 183
Admiral, white **174**
Aeshna affinis **See hawker, southern migrant**
Aeshna mixta **See hawker, migrant**
Agrimonia eupatoria **See agrimony**
Agrimony **134**
Agrostemma githago **See corncockle**
Ajuga chamaepitys **See ground-pine**
Ajuga reptans **See bugle**
Alauda arvensis **See skylark**
Alchemilla filicaulis subspecies vestita **See lady's-mantle, hairy**
Alchemilla mollis See lady's-mantle, soft
Anacamptis morio See orchid, green-winged
Anacamptis pyramidalis **See orchid, pyramidal**
Anax imperator **See dragonfly, emperor**
Andrena florea **See bee, bryony mining**
Andrena hattorfiana **See bee, large scabious mining**
Anemone nemorosa **See anemone, wood**
Anemone, wood **95–97**
Anguis fragilis **See slow-worm**
Anisantha diandra **See brome, great**
Anisantha sterilis **See brome, barren**
Anthemis cotula **See chamomile, stinking**
Anthophora bimaculata **See bee, green-eyed flower**
Anthophora plumipes **See bee, hairy-footed flower**
Anthus pratensis **See pipit, meadow**
Apatura iris **See purple emperor**
Apis mellifera **See bee, honey**
Apple **80**
Aquilegia vulgaris **See columbine, wild**
Archangel, yellow **14, 33, 89, 91**
Arctia caja See tiger, garden
Ardea alba See egret, great white
Argogorytes mystaceus **113–114, 171**
Argynnis paphia **See fritillary, silver-washed**
Arum maculatum **See lords-and-ladies**
Ash **82**
Asilus crabroniformis **See robber-fly, hornet**
Astragalus glycyphyllos **See liquorice, wild**
Atriplex littoralis **See orache, grass-leaved**

Aurochs 217
Avens, wood 216
Badger 95, 111, 189–190, 194–195
Bartsia, red 134, 167
Basil, wild 133
Bedstraw, hedge 133
Bee, box-headed blood 171
Bee, brown-banded carder 134, 173, 178
Bee, bryony mining 168
Bee, gold-tailed melitta 167
Bee, green-eyed flower 171
Bee, hairy-footed flower 172
Bee, honey 165, 169, 180
Bee, ivy 172, 180
Bee, large scabious mining 167–168, 173
Bee, red bartsia 167
Bee, red-tailed mason 171, 233
Bee, sharp-collared furrow 171
Bee, small scissor 167
Bee, small shaggy 171
Beech 25, 48, 82, 93, 115–117, 120
Bellflower, clustered 33, 125, 130–131, 139–140, 144, 167, 217
Bellflower, nettled-leaved 167
Bembecia ichneumoniformis **See moth, six-belted clearwing**
Betonica officinalis See betony
Betony 138
Betula pendula **See birch, silver**
Birch, silver 207
Bird's-foot-trefoil, common 129, 133, 154, 174–175
Bird's-foot-trefoil, greater 154
Bird's-foot-trefoil, narrow-leaved 154
Bitter-cress wavy 151
Bitter-cress, hairy 151
Bitter-cress, large 151
Blackcap 186
Blackstonia perfoliata **See yellow-wort**
Blackthorn 24, 171
Blue, adonis 124, 164–165, 174–176
Blue, chalkhill 124, 165, 174–175, 178
Blue, common 133
Bluebell 33, 79, 89–91, 95–97, 99–100, 104
Bluethroat 213, 215
Boar, wild 49, 67, 90, 96, 138, 216
Bombus humilis **See bee, brown-banded carder**
Bombus hypnorum **See bumblebee, tree**
Bos primigenius See aurochs
Brachytron pratense **See dragonfly, hairy**
Bracken 73, 204
Brome, barren 202

Brome, great	**202**
Brome, interrupted	**154–161, 192, 204, 210, 218**
Brome, rye	**46, 155**
Bromus hordaceus	**See soft-brome**
Bromus interruptus	**See brome, interrupted**
Bromus secalinus	**See brome, rye**
Brooklime	**86, 205**
Broom	**83, 233**
Broomrape, common	**133**
Bryonia dioica	**See bryony, white**
Bryony, white	**163–164, 168**
Buddleia davidii	**See butterfly-bush**
Bufo bufo	**See toad, common**
Bugle	**88**
Bugloss	**46**
Bullfinch	**186**
Bumblebee, tree	**180**
Bunting, corn	**185, 192–193**
Burnet rose	**33, 205**
Burnet, salad	**129, 134**
Burnet-saxifrage	**139**
Butcher's-broom	**91**
Buteo buteo	**See buzzard**
Buttercup, goldilocks	**97, 151–152**
Butterfly-bush	**73–74**
Buzzard	**183, 187**
Calendula officinalis	See marigold, pot
Callistege mi	**See moth, Mother Shipton**
Campanula glomerata	**See bellflower, clustered**
Campanula rotundifolia	**See harebell**
Campanula trachelium	**See bellflower, nettled-leaved**
Cape-gooseberry	**74**
Capreolus capreolus	See deer, roe
Caprimulgus europaeus	See nightjar
Cardamine amara	See bitter-cress, large
Cardamine flexuosa	**See bitter-cress, wavy**
Cardamine hirsuta	See bitter-cress, hairy
Cardamine pratensis	**See cuckooflower**
Carex sylvatica	**See sedge, wood**
Carlina vulgaris	**See thistle, carline**
Carpinus betulus	**See hornbeam**
Carrot, wild	**133, 171**
Castanea sativa	**See chestnut, sweet**
Celandine, lesser	**89, 204**
Centaurea calcitrapa	See star-thistle, red
Centaurea cyanus	See cornflower
Centaurea scabiosa	**See knapweed, greater**
Cephalanthera damasonium	**See helleborine, white**
Cerceris quinquefasciata	**See weevil-wasp, five-banded**
Cerceris rybyensis	**See digger-wasp, ornate-tailed**

Cercyon pygameus	**200**
Certhia familiaris	**See whitethroat**
Cervus elaphus	See deer, red
Chaenorhinum minus	**See toadflax, small**
Chaffweed	29
Chamomile, stinking	**42–44, 46, 55, 60, 68, 175**
Chaser, broad-bodied	**177**
Chelostoma campanularum	**See bee, small scissor**
Chenopodium album	**See fat-hen**
Cherry, wild	**91, 97**
Chestnut, sweet	**82, 100–101, 117, 207**
Chicory	**137**
Chiffchaff	**186**
Chroicocephalus ridibundus	**See gull, black-headed**
Cichorium intybus	**See chicory**
Cinquefoil, creeping	**85, 149**
Cinquefoil, hoary	**137, 211**
Cirsium acaule	**See thistle, dwarf**
Cirsium arvense	**See thistle, creeping**
Cirsium palustre	**See thistle, marsh**
Cirsium vulgare	**See thistle, spear**
Clary, meadow	**13–16, 18–19, 32–33, 140–143, 209, 213–214, 221**
Clary, wild	**14**
Clinopodium vulgare	**See basil, wild**
Clover, red	**134**
Clover, white	**127**
Club-rush, bristle	**204, 205**
Colletes hederae	**See bee, ivy**
Columba livia	See dove, rock
Columba oenas	**See dove, stock**
Columba palumbus	**See woodpigeon**
Columbine, wild	**25, 26, 94, 96**
Conium maculatum	**See hemlock**
Convallaria majalis	**See lily-of-the-valley**
Corncockle	**57–60, 68, 78, 204, 226–227**
Cornflower	37, 78
Cornsalad, broad-fruited	43
Cornsalad, common	**43**
Cornsalad, hairy-fruited	43
Cornsalad, keel-fruited	43
Cornsalad, narrow-fruited	**32, 43, 46–47, 56, 60, 68**
Cornus sanguinea	**See dogwood**
Corvus corax	**See raven**
Corylus avellana	**See hazel**
Cowslip	**136**
Crane's-bill, long-stalked	**67, 204**
Crane's-bill, meadow	**74, 78**
Crassula helmsii	See pigmyweed, New Zealand
Crataegus monogyna	**See hawthorn**

Cuckooflower **89, 106**
Cucullia chamomillae **See moth, chamomile shark**
Cudweed, broad-leaved **18–19, 27, 32, 34, 46, 48, 55–56, 84, 195, 228–229**
Cudweed, common **84**
Cynoglossum officinale **See hound's-tongue**
Cytisus scoparius **See broom**
Dactylorhiza fuchsii **See spotted-orchid, common**
Dactylorhiza traunsteinerioides See marsh-orchid, narrow-leaved (or Pugsley's)
Daisy, oxeye **205**
Dama dama **See deer, fallow**
Dandelion **49, 86, 152**
Daphne laureola **See spurge-laurel**
Darnel 226
Darter, common **177**
Daucus carota **See carrot, wild**
Dead-nettle, henbit **69**
Dead-nettle, red **69**
Deer, fallow **138, 195–197**
Deer, red 195
Deer, roe 138, 195
Digger-wasp, ornate-tailed **171**
Digitalis purpurea **See foxglove**
Dionaea muscipula See fly-trap, Venus
Dog-violet, common **174**
Dogwood **24**
Dormouse, hazel **100, 190**
Dorycera graminum **See phoenix fly**
Dove, rock **188**
Dove, stock **187**
Dove, turtle **164, 185, 193, 213**
Dragonfly, emperor **177**
Dragonfly, hairy **177–178**
Dropwort 24–25, 129
Dryobates minor **See woodpecker, lesser spotted**
Dulse 232
Dutch elm disease **69**
Earth-moss, clustered **185**
Echium vulgare **See viper's-bugloss**
Egret, Great white 213
Eilema griseola **See moth, dingy footman**
Elm, english **69**
Emberiza calandra **See bunting, corn**
Emberiza citrinella **See yellowhammer**
Ephemerum cohaerens **See earth-moss, clustered**
Epipactis helleborine **See helleborine, broad-leaved**
Eptesicus serotinus See bat, serotine
Erinaceus europaeus See hedgehog
Erynnis tages **See skipper, dingy**

Eschscholzia californica See poppy, Californian
Euclidia glyphica **See moth, burnet companion**
Euphorbia exigua **See spurge, dwarf**
Euphrasia nemorosa **See eyebright, common**
Euplagia quadripunctaria **See tiger, Jersey**
Eyebright, common **136, 144–145, 147, 205, 211**
Fagus sylvatica **See beech**
Falco peregrinus **See peregrine**
Falco subbuteo **See hobby**
Falco tinnunculus **See kestrel**
Fat-hen **67**
Fescue, rat's-tail **202**
Ficaria verna **See celandine, lesser**
Fieldfare **186**
Filago germanica **See cudweed, common**
Filago pyramidata **See cudweed, broad-leaved**
Filipendula ulmaria See meadowsweet
Filipendula vulgaris See dropwort
Flax, fairy **133**
Fleawort, marsh 211
Fluellen, round-leaved **46, 61**
Fluellen, sharp-leaved **46**
Fly-trap, Venus 163
Forget-me-not, field **69**
Fox, red **194–195**
Foxglove **83**
Fragaria vesca **See strawberry, wild**
Fraxinus excelsior **See ash**
Fritillary, silver-washed **174–175**
Frog, common **183**
Fumaria densiflora **See fumitory, dense-flowered**
Fumaria officinalis **See fumitory, common**
Fumaria parviflora **See fumitory, fine-leaved**
Fumitory, common **43–44**
Fumitory, dense-flowered **43–44**
Fumitory, fine-leaved **43–46, 56, 211**
Galium album **See bedstraw, hedge**
Garrulus glandarius **See jay**
Genista tinctoria See greenweed, dyer's
Geotrupes spiniger **200**
Geranium columbinum **See crane's-bill, long-stalked**
Geranium pratense **See crane's-bill, meadow**
Geum urbanum **See avens, wood**
Glebionis segetum **See corn marigolds**
Glow-worm **204**
Glyceria fluitans **See sweet-grass, floating**
Gorse **73**
Greenweed, dyer's 138
Gromwell, common **133, 204**
Ground-pine **18–19, 21, 27–28, 46, 67–68, 77, 214,**

(Ground-pine continued) **228**
Groundsel, heath **83**
Guelder-rose **91**
Gull, black-headed **184**
Gull, Mediterranean **184**
Hare's-tail grass **137**
Harebell **33, 124, 129, 152–153**
Hawkbit, autumn **216**
Hawker, migrant **177**
Hawker, southern migrant **181**
Hawkweed, autumn **86, 152**
Hawthorn **207**
Hazel **97, 101, 207, 223**
Hedera helix **See ivy, common**
Hedgehog **189**
Helianthemum nummularium **See rock-rose, common**
Helicella itala See snail, heath
Helleborine, broad-leaved **104, 119**
Helleborine, white **106, 115–117, 120**
Helminthotheca echioides **See oxtongue, bristly**
Hemlock **226**
Heracleum sphondylium **See hogweed**
Hieracium sabaudum **See hawkweed, autumn**
Hippocrepis comosa **See vetch, horseshoe**
Hobby **177, 187**
Hogweed **171**
Holly **207**
Homo heidelbergensis 62
Homo neanderthalensis 62
Hornbeam **82, 93, 101, 115**
Hound's-tongue **137**
Hyacinthoides non-scripta **See bluebell**
Hypericum androsaemum **See tutsan**
Hypericum humifusum **See St John's-wort, trailing**
Hypericum tetrapterum **See St John's-wort, square-stalked**
Ichthyaetus melanocephalus **See gull, Mediterranean**
Ilex aquifolium **See holly**
Iris foetidissima **See iris, stinking**
Iris, stinking **94**
Isolepis setacea **See club-rush, bristle**
Ivy, common **73, 180**
Jacobaea erucifolia **See ragwort, hoary**
Jacobaea vulgaris **See ragwort, common**
Jay **129**
Kestrel **187**
Kickxia elatine **See fluellen, sharp-leaved**
Kickxia spuria **See fluellen, round-leaved**
Kite, red **187–188**
Knapweed, greater **123–125, 129, 134, 204**
Knautia arvensis **See scabious, field**

Knotweed, japanese **73–74**
Lady's-mantle, hairy **88–89, 151**
Lady's-mantle, soft 88
Lagurus ovatus **See grass, hare's-tail**
Lamiastrum galeobdolon **See archangel, yellow**
Lamium amplexicaule **See dead-nettle, henbit**
Lamium purpureum **See dead-nettle, red**
Lampyris noctiluca **See glow-worm**
Lasioglossum malachurum **See bee, sharp-collared furrow**
Lathraea squamaria **See toothwort**
Lathyrus nissolia **See vetchling, grass**
Legousia hybrida **See Venus's-looking-glass**
Leucanthemum vulgare **See daisy, oxeye**
Libellula depressa **See chaser, broad-bodied**
Lily-of-the-valley **86–87, 91**
Limax cinereoniger **See slug, ash-black**
Limenitis camilla **See admiral, white**
Linum catharticum **See flax, fairy**
Liquorice, wild **132–134**
Lithospermum officinale **See gromwell, common**
Lizard, viviparous (or common) lizard **183**
Lolium perenne **See rye-grass, perennial**
Lolium temulentum See darnel
Lords-and-ladies **106**
Lotus corniculatus **See bird's-foot-trefoil, common**
Lotus pedunculatus **See bird's-foot-trefoil, greater**
Lotus tenuis See bird's-foot-trefoil, narrow-leaved
Love-in-a-mist 163
Luscinia svecica See bluethroat
Luzula forsteri **See wood-rush, southern**
Lycopus arvensis **See bugloss**
Lysimachia arvensis **See pimpernel, scarlet**
Lysimachia foemina **See pimpernel, blue**
Lysimachia minima **See chaffweed**
Lytta vesicatoria **See Spanish fly beetle**
Madder, field **67–68**
Mallow, rough **15–19, 21, 27–28, 32, 75–77, 194, 214, 221**
Malus domestica **See apple**
Malva setigera **See mallow, rough**
Maple, field **24, 82, 97, 101**
Marigold, corn 78
Marigold, pot 163
Marjoram, wild **133**
Marsh-orchid, narrow-leaved (or Pugsley's) 153
Matricaria chamomilla **See mayweed, scented**
Mayweed, scented **43**
Mayweed, scentless **43, 175**
Meadow-grass, wood **91**
Meadowsweet 25, 233

Meles meles See badger
Melica uniflora See melick, wood
Melick, wood 91
Melitta haemorrhoidalis See bee, gold-tailed melitta
Melitta tricincta See bee, red bartsia
Mentha arvensis See mint, corn
Mercurialis annua See mercury, annual
Mercurialis perennis See mercury, dog's
Mercury, annual 41
Mercury, dog's 89, 94–95, 99
Micromys minutus See mouse, harvest
Milium effusum See millet, wood
Milkwort, common 138
Millet, wood 216
Milvus milvus See kite, red
Mint, corn 46, 48
Moth, black-veined 210
Moth, burnet companion 174
Moth, chamomile shark 175
Moth, dingy footman 174
Moth, garden tiger 181
Moth, jersey tiger 180–181
Moth, Mother Shipton 174
Moth, oak processionary 181–182
Moth, rosy-striped knot-horn 174
Moth, six-belted clearwing 134
Moth, six-spot burnet 174
Mouse, harvest 190–191, 193, 205
Muscardinus avellanarius See dormouse, hazel
Myosotis arvensis See forget-me-not, field
Natrix helvetica See snake, grass
Neanderthal humans 62
Neottia nidus-avis See orchid, bird's-nest
Neottia ovata See twayblade, common
Newt, great crested 183
Nigella damascena See love-in-a-mist
Nightjar 186
Nuthatch 186
Oak, pedunculate 24, 82, 149, 207
Oak, sessile 82, 91, 149
Odontites vernus See bartsia, red
Oncocera semirubella See moth, rosy-striped knot-horn
Onobrychis viciifolia See sainfoin
Ononis repens See restharrow
Ophiostoma sp. See Dutch elm disease
Ophrys apifera See orchid, bee
Ophrys insectifera See orchid, fly
Orache, grass-leaved 137
Orchid, bee 102, 118, 120, 122, 134
Orchid, bird's-nest 117, 120

Orchid, early-purple	**33, 100, 104–106**
Orchid, fly	**32, 106, 113–115, 117–118, 120, 122, 171, 216**
Orchid, green-winged	138
Orchid, lady	**32–33, 94, 106, 109–113, 115, 117, 120, 194, 196, 216**
Orchid, man	**21, 33, 106–109, 115, 117, 120, 129, 131, 138, 205, 217**
Orchid, pyramidal	**102, 119**
Orchis anthropophora	**See orchid, man**
Orchis mascula	**See orchid, early-purple**
Orchis purpurea	**See orchid, lady**
Origanum vulgare	**See marjoram, wild**
Orobanche minor	**See broomrape, common**
Oryctolagus cuniculus	**See rabbit**
Osmia bicolor	**See bee, red-tailed mason**
Owl, tawny	**186**
Oxalis acetosella	See wood-sorrel
Oxtongue, bristly	**202**
Oxtongue, hawkweed	**202**
Palmaria palmata	See dulse
Pansy, Altaian	148
Pansy, field	**51, 147–149**
Pansy, garden	148
Pansy, mountain	148
Pansy, wild	148–149
Panurgus calcaratus	**See bee, small shaggy**
Papaver dubium	**See poppy, long-headed**
Papaver rhoeas	**See poppy, common**
Papaver somniferum	**See poppy, opium**
Passer domesticus	**See sparrow, house**
Pepper-saxifrage	138
Peregrine	**187–189**
Persicaria hydropiper	**See water-pepper**
Phoenix fly	**178**
Phylloscopus collybita	**See chiffchaff**
Phylloscopus trochilus	**See warbler, willow**
Physalis peruviana	**See cape-gooseberry**
Picris hieracioides	**See oxtongue, hawkweed**
Pigeon, feral	**188**
Pigmyweed, New Zealand	73
Pimpernel, blue	**28, 46, 50–53, 55**
Pimpernel, scarlet	**41, 48, 50–52, 223**
Pimpinella saxifraga	**See burnet-saxifrage**
Pine, scots	**74**
Pinus sylvestris	**See pine, Scots**
Pipit, meadow	**186**
Plantago major	**See plantain, greater**
Plantain, greater	**235**
Poa nemoralis	**See meadow-grass, wood**

Podarcis muralis	**See lizard, wall**
Poecile montanus	See tit, willow
Poecile palustris	**See tit, marsh**
Polygala vulgaris	**See milkwort, common**
Polyommatus bellargus	**See blue, adonis**
Polyommatus coridon	**See blue, chalkhill**
Polyommatus icarus	**See blue, common**
Poppy, Californian	163
Poppy, common	**37, 39–41, 58, 61, 68, 78–79, 160, 192, 218, 222–225, 229, 234**
Poppy, long-headed	**39–41, 61**
Poppy, opium	**37, 39–40, 68, 224**
Poppy, prickly	**39–41, 46, 61, 68, 136, 202, 225**
Poppy, rough	**39–41, 68, 136**
Potentilla anglica	**See tormentil, trailing**
Potentilla argentea	**See cinquefoil, hoary**
Potentilla erecta	See tormentil
Potentilla reptans	**See cinquefoil, creeping**
Potentilla sterilis	**See strawberry, barren**
Poterium sanguisorba	**See burnet, salad**
Primrose	**88, 204, 209**
Primula veris	**See cowslip**
Primula vulgaris	**See primrose**
Privet, wild	**24**
Prunus avium	**See cherry, wild**
Prunus spinosa	**See blackthorn**
Pteridium aquilinum	**See bracken**
Purple emperor	**211, 213**
Pyrgus malvae	**See skipper, grizzled**
Pyrrhula pyrrhula	**See bullfinch**
Quercus petraea	**See oak, sessile**
Quercus robur	**See oak, pedunculate**
Rabbit	**113, 115, 129, 189, 214, 216**
Ragwort, common	**166**
Ragwort, hoary	**134**
Rana temporaria	**See frog, common**
Ranunculus auricomus	**See buttercup, goldilocks**
Ranunculus baudotii	**See water-crowfoot, brackish**
Raven	**188, 209, 234**
Redpoll, lesser	**186**
Redwing	**186**
Restharrow	**134**
Reynoutria japonica	**See knotweed, Japanese**
Rhinanthus minor	**See yellow-rattle**
Rhododendron	**73**
Rhododendron spp.	**See rhododendron**
Robber-fly, hornet	**200–201**
Rock-rose, common	**124, 129, 204**
Roemeria argemone	**See poppy, prickly**
Roemeria hispida	**See poppy, rough**

Rosa micrantha **See sweet-briar, small-flowered**
Rosa spinosissima **See rose, burnet**
Ruscus aculeatus **See butcher's-broom**
Rye-grass, perennial **127, 226**
Sainfoin 158–161
Salix caprea **See willow, goat**
Salvia pratensis **See clary, meadow**
Salvia verbenaca See clary, wild
Saxicola rubicola **See stonechat**
Scabiosa columbaria **See scabious, small**
Scabious, field **168**
Scabious, small **7, 134, 168, 204**
Scolopax rusticola **See woodcock**
Scorzoneroides autumnalis **See hawkbit, autumn**
Sedge, wood **91**
Senecio sylvaticus **See groundsel, heath**
Serotine bat 200
Shelduck **189**
Sherardia arvensis **See madder, field**
Silaum silaus See pepper-saxifrage
Siona lineata See moth, black-veined
Sisyrinchium striatum **See yellow-eyed-grass, pale**
Sitta europaea **See nuthatch**
Skipper, dingy **134, 174**
Skipper, grizzled **133–134, 174–175**
Skylark **185–186, 192–193**
Slow-worm **183**
Slug, ash-black **91**
Small toadflax **46**
Snail, heath 66
Snake, grass **183**
Soft-brome **155, 202**
Sonchus arvensis **See sowthistle, perennial**
Sorbus aria: **See whitebeam, common**
Sowthistle, perennial **202**
Spanish fly beetle **180**
Sparrow, house **192–193, 218**
Sparrowhawk **187**
Speedwell, wood **86–87, 91**
Spergula arvensis See spurrey, corn
Sphecodes monilicornis **See bee, box-headed blood**
Spotted-orchid, common **102, 118, 153**
Spurge, dwarf **46**
Spurge-laurel **94**
Spurrey, corn 217
St John's-wort, square-stalked **86**
St John's-wort, trailing **84–85, 87**
Star-thistle, red 214–215
Stegosaurus 1
Stonechat **186**

Strawberry, barren	**99**
Strawberry, wild	**129, 134, 174**
Streptopelia turtur	**See dove, turtle**
Strix aluco	**See owl, tawny**
Sus scrofa	See boar, wild
Sweet-briar, small-flowered	**205–206**
Sweet-grass, floating	**137**
Sylvia atricapilla	**See blackcap**
Sylvia borin	**See warbler, garden**
Sympetrum striolatum	**See darter, common**
Tadorna tadorna	**See shelduck**
Taraxacum spp	**See dandelion**
Tephroseris palustris	See fleawort, marsh
Thaumetopoea processionea	**See moth, oak processionary**
Thistle, carline	**136**
Thistle, creeping	**166**
Thistle, dwarf	**124, 134, 139**
Thistle, marsh	**86**
Thistle, spear	**166**
Thyme, Breckland	46
Thyme, large	**124, 134, 205**
Thyme, wild	48
Thymus drucei	See thyme, wild
Thymus pulegioides	**See thyme, large**
Thymus serpyllum	See thyme, Breckland
Tit, marsh	**186**
Tit, willow	213, 215
Toad, common	**183**
Toothwort	**97–98**
Tormentil	85, 149
Tormentil, trailing	**84–85, 149**
Triceratops	1
Trifolium pratense	**See red clover**
Trifolium repens	**See clover, white**
Tripleurospermum inodorum	**See mayweed, scentless**
Triturus cristatus	**See newt, great crested**
Turdus iliacus	**See redwing**
Turdus pilaris	**See fieldfare**
Tutsan	**74, 91**
Twayblade, common	**117, 119–120**
Tyrannosaurus	1
Ulex europaeus	**See gorse**
Ulmus procera	**See elm, English**
Valerianella carinata	See cornsalad, keel-fruited
Valerianella dentata	**See cornsalad, narrow-fruited**
Valerianella eriocarpa	See cornsalad, hairy-fruited
Valerianella locusta	**See cornsalad, common**
Valerianella rimosa	See cornsalad, broad-fruited
Venus's-looking-glass	**21, 28, 32, 41**
Veronica beccabunga	**See brooklime**

Veronica montana	**See speedwell, wood**
Vetch, horseshoe	**25, 124, 129, 139–140, 164**
Vetchling, grass	**132–133, 205**
Viburnum lantana	**See wayfaring-tree**
Viburnum opulus	**See guelder-rose**
Viola altaia	See pansy, Altaian
Viola arvensis	**See pansy, field**
Viola hirta	**See violet, hairy**
Viola lutea	See pansy, mountain
Viola riviniana	**See dog-violet, common**
Viola tricolor	See pansy, wild
Viola* x *contempta	**148**
Viola x *wittrockiana*	See pansy, garden
Violet, hairy	**133**
Viper's-bugloss	**46, 202–203**
Vipera berus	See adder
Volucella bombylans	**179**
Vulpes vulpes	**See fox, red**
Vulpia myuros	**See fescue, rat's-tail**
Wall lizard	183
Warbler, garden	**186**
Warbler, willow	**184**
Water-crowfoot, brackish	**137, 211**
Water-pepper	**205**
Wayfaring-tree	**24**
Weevil-wasp, five-banded	**167, 178**
Whitebeam, common	**24**
Whitethroat	**186, 204**
Willow, goat	**207**
Woodcock	**184–186**
Woodpecker, lesser spotted	**186**
Woodpigeon	**187**
Wood-rush, southern	**91–92**
Wood-sorrel	**216**
Yarrow	**171**
Yellow-eyed-grass, pale	**74**
Yellowhammer	**185, 191, 193**
Yellow-rattle	**133–135, 211**
Yellow-wort	**124–125, 133**
Zootoca vivipara	**See lizard, viviparous (or common)**
Zygaena filipendulae	**See moth, six-spot burnet**

General Index

Atkinson, Chris	147
Attenborough, David	22
Back from the Brink project	27–28
Bailey, Charles	18
Bellamy, David	28, 31
Bennett, George	11
Bennett, Ros	31
Bligh, Ivo, 8th Earl of Darnley	11
Bligh, John, 1st Earl of Darnley	11
Bligh, John, 6th Earl of Darnley	220
Botanical Society of Britain and Ireland	See BSBI
Botanical Society of London	16
Broad Oak Wood	9
Brockles Field, Ranscombe	8, 21, 33, 109, 118–119, 125, 131–132, 134–136, 138–140, 144–145, 173–174, 200, 205
Bronze Age	66–68
Brooke, William	196–197
Brown, Mervyn	23
BSBI	16, 73, 146–147, 159
Bush Valley	9
Byfield, Andy	29
Cement industry	5
Channel Tunnel Rail Link (HS1)	7–8, 9, 29, 65, 74, 99
Chatham Historic Dockyard	5
Chatham	iv, 4–6, 9, 82, 215, 219
Childish, Billy	8
Chris Valdus	30
Cobbett, William	194
Cobham	11–13, 16, 18, 101, 157, 219–221
Cobham Ashenbank Management Scheme (CAMS)	29
Cobham Field, Ranscombe	55
Cobham Hall	11–12, 20, 93, 196–197, 220–221, 244
Cobham Park	11, 15, 29, 106
Cobham Woods	8–9, 15, 18, 27, 29, 55, 62, 74, 80, 83–86, 89, 91–93, 99, 101, 104, 110, 117, 129, 137, 141, 149, 151–152, 154, 171, 174–175, 183–184, 195, 207, 219–221
Cogger's Valley, Ranscombe	9
Coldrum Stones	65
Costley, Joe	30, 138
Court Lodge Farm	9, 12
Cuxton Countryside Group	129
Cuxton Weedy Field	9, 28, 32, 35
Darland Banks	176, 215

de Cobham, Henry 196–197
Dickens, Charles 219–221, 234
Druce, George Claridge 18, 159
Dugmore, Arthur 224
Dukes of Richmond 11
Earls of Darnley 11–12, 220
East Thames Corridor 167–168, 178, 180, 200
Fertile Crescent 60–61, 67, 160, 192
Foulds, Bob 24
Gerard, John 13
Gillingham 4, 13, 25, 215
Great Lines 4, 215
Great Wood, Ranscombe 9, 19
Hall, Martin 29
Henslow, Rev. Prof. John Stevens 16–18
Heritage Lottery Fund 33, 35
In Flanders Fields (poem) 223
Iron Age 66–68, 73, 81
Irvine, Alexander 18
James Taylor Quartet 8
Johnson, Thomas 13–14, 33, 221
Kit's Coty House 65
Kitchen Field, Ranscombe 9–10, 19, 21, 27–28, 32, 35, 42, 50–51, 55, 57–58, 71, 77, 106–109, 115, 118, 129, 137–138, 174–175, 183, 202, 206, 208, 222, 224, 233
Knepp 211, 213, 215
Knock Field, Ranscombe 9, 128
Last Orders (book) 5
Law, Ken, 23
Lemon, Stephen 185
Lingham, Andrew 33, 198, 200–201
Little Bottoms Field, Ranscombe 9
London, Chatham & Dover Railway Co. 6, 93
Longhoes Field, Ranscombe 7–8, 42, 56, 202–203, 205
Lousley, J. E. (Ted) 18–19, 22, 27
M2 motorway 1, 6, 9, 189
May, David 62–65, 70, 71
McCrae, John 223–224
Medieval Period 7, 69–72, 81, 93, 221, 234
Medway Megaliths 65
Medway Towns iv, 1, 4–7, 9, 20, 25, 176, 188, 219
Merrals Shaw, Ranscombe 8, 81, 104, 137, 177
Monet 224
Morley, Helen 30–33, 56, 141
National Plant Monitoring Scheme 210
Natural England 28, 131
Neolithic Period 65–67, 81, 100, 122–123, 154, 191, 193
Nether Great Bottoms, Ranscombe 9
North Downs Way 8, 24, 69, 219

North Kent Wildlife Preservation Soc. 23, 25
Pamplin, William 18
Plukenet, Leonard 14, 28
Poppies and Butterflies (painting) 225
Prebbles Field, Ranscombe 81
Pye, William 12
Queendown Warren 176
Ranscombe Studios 8
Ranscombe Wildlife Study Group 190, 200
Repton, Humphry 11
River Medway 1, 4–6, 19, 65, 178
Rochester 4–7, 12–14, 16, 20, 69, 219–221
Rochester and District Natural History Society (Rochester 'Nats') 23, 25, 55
Romans 4, 20, 68–70, 219, 221
Royal African Company 12
Rumsey, Fred 156–157
Shakespeare, William 225, 227, 234
Sir Joseph Williamson's Mathematical School 12
Skipp, Stephanie 114–115
Smith, Phillip Morgans 156
Society of Apothecaries 13
South Eastern Railway Company 6
Southern Valley, Ranscombe 2, 9, 19, 21–22, 24
Springate, Simon 199
Steere, David 117
Steve Kirk 190
Stewart, Katherine 11, 13
Stogarts and Drapers Wood, Ranscombe 9
Swanscombe Peninsula 178
Sweeney, Ben 31, 33, 141, 207
Symons, Rev. Jelinger 16
The Mabinogion 232–234
The Pickwick Papers (book) 220–221
The Poppy Field near Argenteuil (painting) 224
The Seven Poor Travellers (story) 221
Thornton, Kevin 193
Town Road 80, 85, 93, 152, 219–221
Troops Going over the Top, First World War (painting) 224
Trosley Country Park 214
Tuson, Dan 125
Van Gogh 225
Volunteers iii, 24, 30–32, 35, 62–63, 109, 129, 147, 157, 169, 185, 188, 190, 200, 208, 211
Watling Street 4–6, 9, 14, 20, 69, 219
Weald-Artois anticline, 1–3, 9
White, Gilbert 191

Williamson, Sir Joseph 11–13, 24
Worrall, Pam 190